Leckie×Leckie
Scotland's leading educational publishers

Active LEA
Curriculum for Excellence

Third Level
GEOGRAPHY

Rob Hands

CONTENTS

GIS AND MAPWORK SKILLS

EARTH FORCES

GLACIATED LANDSCAPES

LANDSCAPES AND RIVERS

COASTAL LANDSCAPES

CONTENTS

INTRODUCTION

This book is about learning, and active learning in particular. Its aim is to support the Curriculum for Excellence in Geography, so it is an interpretation of the **experiences and outcomes** relating to geographical studies and skills based on the specific statements given in People, Place and Environment. The book is rooted in the Third Level but recognises that many students in S1, and progressively in S2 and S3 need to move on and there is a need to raise the bar for them. This means that many exercises in the book take on and explore Fourth Level outcomes and experiences.

Geography departments have an existing range of textbooks and cover a wide range of topics, so although this book might form a good companion to a whole course, it does not claim to be a course book in the traditional sense. Instead it provides a series of content examples but sets them in the context of active learning, using tried and tested activities and laying these out in detail. In this way the methods and skills being developed should be the dominant feature, and their introduction into various units of the book should allow them to be used as templates where new content can be substituted or superimposed on the framework provided to either replace it or to further develop the skills introduced: therefore supporting progression and developing competence and independence in learning.

Another feature of the book is the reference on a regular basis to **co-operative learning**. It is the belief of the author that collaborative learning is a powerful tool in the classroom and is an obvious way to make learning active and enjoyable for students. Co-operative learning is predominantly teacher led but students and parents can take the collaborative structures it provides to work on things at home, or in independent groups to enhance their learning, provided the instructions are clear.

The book attempts to encourage every student to become an independent researcher able to follow his or her ideas and present them to an audience of other group members, fellow students, teachers and parents and other interested groups of people.

Third Level Active Geography also contributes to Literacy, Numeracy, Health and Wellbeing, and, through its development of map skills, to Outdoor Learning. Care has been taken to ensure that, through the investigative approach, a considerable contribution is made towards the delivery of Literacy experiences and outcomes.

Finally, this book is about learning, it is about getting students to be active. It is about stretching students by providing interesting and stimulating learning contexts and, in some cases, different ways of approaching the study of a topic. It recognises that students do not learn in isolation but can work in collaborative teams to help each other and support each other's learning. It recognises that students, with the support of teachers and parents, can further develop the scope and complexity of what they learn away from the Geography class as well.

CO-OPERATIVE LEARNING

Co-operative learning is used widely in Scottish schools and there has been a large investment of time and resources to train teachers in its use. It is regarded as one of the most effective ways for students to learn.

As a teaching and learning strategy it gets students into small teams with differing abilities and sets up a variety of **active learning** situations. A lot of co-operative learning is about communication between group members and the development of the social skills that promote good communication. This strategy encourages students to take responsibility for their own learning but also for the learning of others in their group, such that if one succeeds on a task all succeed. If used effectively, co-operative learning establishes a growing and justified confidence among students and encourages engagement in the work of the class. It also improves the depth of learning and opens up the learning process, as children and young adults recognise that they can be successful learners and they can make a contribution to any team effort.

Research shows that co-operative learning:

- improves pupil learning and academic achievement
- enriches the students' learning experience
- develops student oral communication skills
- increases the retention and development of knowledge and skills
- promotes positive relationships in class
- promotes student self-esteem
- develops student social skills
- improves classroom behaviour

HOW DOES CO-OPERATIVE LEARNING WORK?

It is beyond the scope of this book to satisfactorily answer this question. But, in brief, it is possible to give a flavour of how this learning system works. Firstly it moves away from more traditional competitive individual learning. It is different from traditional group work because of its highly structured nature and its use of positive interdependence and individual responsibility. It also places emphasis on social goals as well as academic learning goals. It is believed that the growing development of social and interpersonal skills plays a major part in developing learning achievement.

CO-OPERATIVE OPERATIONS

This book lays out a few co-operative learning operations in a Geography or investigation context. The methods and activities used in co-operative learning are wide ranging, and it is way beyond the scope of this book to do little more than scratch the surface, but the book does provide a number of exemplars that can be used directly and can form templates for similar activities. In a class teaching context however, they will need to be further developed to suit the way an individual class works with collaborative learning. The activities mostly used are pair-share and talking around the table, taking turns and jigsaw operations. There is an example of the use of dotmocracy and team game tournament but all of these operations are given in relatively simple forms to increase the likelihood of them being useful. All the operations will benefit from enhanced planning before they are used. Debate and structured discussion are also used, but again it will be up to the director of the study programme to adapt this and to build in his or her own improvements.

INTRODUCTION

CAN COLLABORATIVE/CO-OPERATIVE LEARNING BE DONE AT HOME?

Yes it can and this is increasingly the case through the use of the internet and digital media. There are a number of books that detail co-operative learning 'structures' such as Kagan's *Co-operative Learning*. There is a range of lesson plans and activities pre-prepared on the net among the teaching and learning websites used by professionals.

Any parent interested in supporting their child with co-operative learning might find these sources a good starting point.

HOW COULD CO-OPERATIVE LEARNING BE ORGANISED AT HOME?

Co-operative learning is based on small groups, ideally groups of two, three or four. The group needs to be bonded into a strong team and a number of social activities will achieve this, such as learning games, pair-share and round the table discussions. All this encourages face-to-face interaction and establishes team working. The normal pattern of the isolated student working alone on an assignment could be replaced at appropriate times by a home study group.

Social networking sites can be used positively to allow contact between groups of students. Simple teleconferencing, for example using Skype, would create an enhanced form of positive interaction for students unable to work in each others' homes. Email can be used for contact and sharing ideas which would key into the literacy agenda.

SIMPLE CO-OPERATIVE LEARNING OPERATIONS

1. Identify the academic goals (specific learning intention). For example, in a study of the effects of glaciation the concepts of plucking and abrasion as ways in which glaciers erode are key parts of assessment and examinations.
2. Identify the social goals (social skills) that will enhance learning. For example, listening to the point of view of another person; or criticising another person's viewpoint in a constructive way.
3. Explain the way these goals can be achieved.
4. Create and bond the team.
5. Set up the learning environment for the task.
6. Set the learning task and allocate roles or develop a path through it to be followed.
7. Build in positive interdependence and individual responsibility. For example, devise strategies that check on the group and make them accountable.
8. Ensure that students always have help available from the group or a supporter.
9. Complete the task.
10. Review the way things went, emphasising what went well and discuss how things could be made better another time.

SUPPORTING LEARNING: TRAFFIC LIGHTS

For any unit it is possible to use a traffic light system that allows students to self-evaluate their performance. The template at the top of the next page is a suggestion of how this might be produced in a straightforward form.

CRASH SHIP 1

S.O.S. from crash ship 1: Mayday mayday! The navigation system is out, we are going in. We will land on island ...(message incomplete)

Suggested rescue strategy: Use a RANDOM SEARCH, choose any island numbered 1 – 50.

CRASH SHIP 2

Mayday mayday! We've been hit, damage to the mapping computer, cannot give an accurate fix. We're going over a skull-like island, heading for an X-shaped island. We'll land along that line please hurry.

Suggested rescue strategy: do a line search and choose a number along the line of flight.

CRASH SHIP 3

Mayday mayday! Code red! We are going to touch down on an island in the south-west sector, north-east of Skull Island. Please hurry, we seem to be sinking into the marsh.

Suggested rescue strategy: Use the directions given to find the island.

CRASH SHIPS 4, 5, 6 & 7

Message from the Air Sea Rescue Co-ordinator: Three crashed ships have reported their location with simple latitude and longitude coordinates as follows: **Ship 4** 2°N 1°E, **Ship 5** 1°N 2°W, **Ship 6** 2°S 0°W, **Ship 7** 2°N 3°W

CRASH SHIPS 8, 9 & 10

Message from the Air Sea Rescue Co-ordinator: Three crashed ships have reported their location with simple latitude and longitude coordinates as follows: **Ship 8** 1°20'N 1°20'E, **Ship 9** 0°40'N 0°30'W, **Ship 10** 1°50'S 0°25'E

CRASH SHIP RESCUE LOG

Calculate your score and answer the debriefing questions below.

Crash ship number	Latitude	Longitude	Island number
1.	N/A	N/A	Teacher chooses
2.	N/A	N/A	Teacher chooses
3.	N/A	N/A	Teacher chooses
4.			
5.			
6.			
7.			
8.			
9.			
10.			

1. Why is it unlikely that you found the island that crash ship 1 landed on?
2. What are the chances of locating the correct island for crash ship 1?
3. Why is it easier to locate crash ship 2? (Hint: count the total number of islands.)
4. What are the chances of finding it? (Hint: count the number of possible landing sites.)
5. What made crash ship 3 the easiest so far to locate?
6. Why is simple latitude and longitude an effective way of finding things on the map?
7. Why would we need to use more complex latitude and longitude coordinates involving minutes as well as whole degrees?

See page 124 for a cooperative learning task relating to this activity.

ORDNANCE SURVEY MAPS

Learning intentions: in this section you will learn to use four- and six-figure coordinates to find and locate places on Ordnance Survey (OS) maps.

One of the most common maps you will use in geography is the Ordnance Survey map. This is produced in Britain by the national mapping agency called the Ordnance Survey, which is based in Southampton in southern England. They produce all kinds of maps that cover the whole of the UK. The web link below will allow you to find a map of anywhere you are keen to look up. You might want to see if you can find your home or your school. www.ordnancesurvey.co.uk/oswebsite/getamap/

Knowing how to use OS maps is one of the key skills of geography and it is one of the important life skills you will learn at school. Knowing how to access and use OS maps accurately is useful in all sorts of ways.

BASIC COORDINATES

To find square C4 (the smiley face), run a finger along the bottom of the map and find column C. Next run another finger up the side of the map and find row number 4. Run your finger up column C and run your other finger across row 4 – where your two fingers meet is given by coordinates C4.

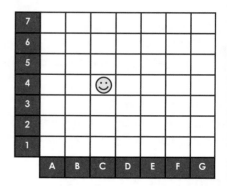

COORDINATES TASK

Draw a grid like the one above, using a pencil and ruler to accurately create the grid squares. Follow the table below to colour the squares in your grid and create the **shape of success.**

Square coordinates	Colour	Square coordinates	Colour
B4	Orange	C3	Yellow
D4	Green	E5	Light blue
G7	Purple	A5	Red
F6	Dark blue	G1	Black

Map A

COORDINATES AND ORDNANCE SURVEY SYMBOLS

Use the map (Map A) on the left and your knowledge of coordinates to complete the table below.

Symbol	Coordinates
	G7
Coniferous forest	
	D3
	B4
'A' class road	
Motorway	

24

STRAIGHTFORWARD COORDINATES: FOUR-FIGURE REFERENCES

Map B

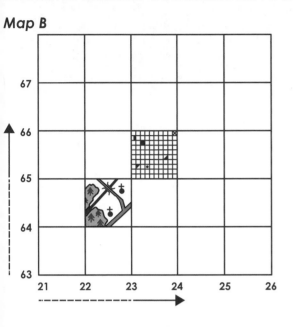

The OS uses a different map referencing system on its maps to the ones we have used so far. It uses the National Grid to identify particular squares on any Ordnance Survey map, which lets us identify any square kilometre in the UK by its unique reference on the Ordnance Survey grid. This is done using a series of letters and numbers but for the most part we will be using a single map sheet so it is only necessary to use the grid numbers on the map we are dealing with. This allows us to identify a kilometre square on the map by using a four-figure reference.

On Map B, a system of numbers is used to locate an individual chequered square (2365). Go along the bottom of the map, reading from left to right until you reach line 23, which marks the beginning of column 23. Next, as with other coordinate maps we have used, go up the side of the map until you reach line 65, which runs along the bottom of row 65. If you run your fingers up the column (23) and across the row (65) you will arrive at grid square 2365 (the chequered square).

GET ACTIVE — MAP REFERENCES TASK

1. Copy the table below and input the data by referring to the map below.

Place	Four-figure grid reference
Kat Hall	
Jean Halt	
River Island	
Spot height 5m	
	1941
Rabton farm	
Monument	
	1838
Jackton Station	
The Quarry	

JACKTON MAP

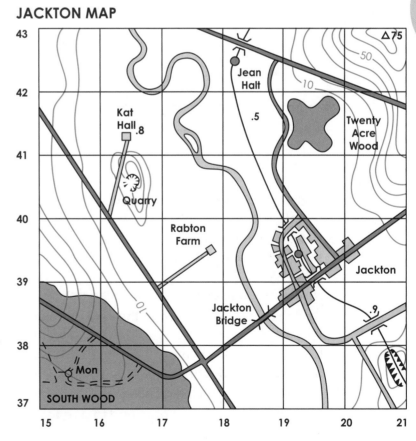

2. Using grid references to identify places, describe a journey from Kat Hall to Jean Halt by road and railway. For example, *Leaving Kat Hall in 1641, head south-west to the road junction at 1640,* and so on.

3. If you were asked where the station is on the map, why would it be difficult to give a grid reference that gives its location?

4. Give the grid reference for the highest land on Jackton Map.

MORE ACCURATE COORDINATES: SIX-FIGURE REFERENCES

Using **four-figure** grid references, we can locate a square and all the things that are in it. Map C (below) is an enlargement of part of Map B on the previous page. You can locate the coloured mapped square and the chequered square along with their reference numbers.

Map C

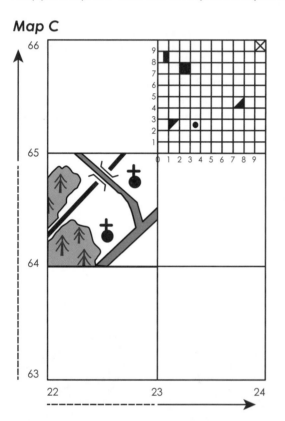

Four-figure references are useful but they leave us with a problem. For example, you might say that in square 2264 there is a railway, and the railway symbol in the square confirms this. But if you were to arrange a meeting at the church in 2264 there is a problem in knowing exactly which church of the two present in the grid square you mean.

It is sometimes crucially important to pinpoint an exact position on the map, for example to direct help for a rescue in the mountains. To do this we need to use a **six-figure** grid reference. To locate the black spot in the chequered square 2365 we must divide the grid square up into tenths along the bottom and up the side (see Map C). You can estimate the number of tenths along and up the way so that they can be slotted into the appropriate place in the original four figure grid reference (2365). Now follow the steps below to pinpoint the black spot with a six-figure reference.

1. Find the grid line before the black spot, as you move left to right along the bottom of the map (line **23**).
2. Now count the number of tenths along, still moving left to right across **column 23**. (Notice the first tenth is numbered 0.) The black spot is on the **three-tenths** line (see Map C).
3. Next write the **three figures** down (the column reference and the tenths, **233**).
4. Next move up the way along the left-hand side of the map as far as line **65** to locate the **row**.
5. Now calculate the number of tenths up the way until you make contact with the black spot. Remember, the first division is 0 so the spot is on the **two-tenths** line.
6. Next write down the three figures: the row reference and the tenths (**652**).
7. Combine the two figures obtained to give the unique six-figure reference for the black spot – **233652**.
8. Work out the references for the other shapes in the chequered square.

26

GET ACTIVE   **SIX–FIGURE REFERENCES**

1. On Map C, give a six-figure grid reference for the most northerly church.

2. What can be found at 225648?

3. Return to Jackton Map (on page 25) and redraw the original table. Add another column to the table labelled 'six-figure references'. Calculate the six-figure references for the places on the table.

Place	Four-figure grid ref.	Six-figure grid ref.
Kat Hall		
Jean Halt		
River Island		
Spot height 5m		
	1941	
Rabton farm		
Monument		
	1838	
Jackton Station		
The Quarry		

TOP TIP
Remember the first tenth is zero so you always slightly overestimate measurement of tenths in a grid reference. You are allowed to be one digit out in your estimation.

MAKE THE LINK

Grid references can be useful in other subjects where OS maps are used. In Biology they might be needed to find sites on a field study visit. In History they could be used to locate historic features such as battlefields, castles and forts or ancient monuments.

DID YOU KNOW?

Mountain rescue teams rely on using exact six-figure grid references to locate accident victims out on the hills. They are also used by the military to set targets and to identify the position of both friends and enemies. The correct grid reference can be a matter of life or death.

Mountain rescue teams using a helicopter to rescue a stranded hill walker

27

ORDNANCE SURVEY MAP WORK

CONTOURS ON OS MAPS

Contours on OS maps show the shape of the land (**relief**) and give the height of the land above sea level. A contour is a line on the map joining places of equal height.

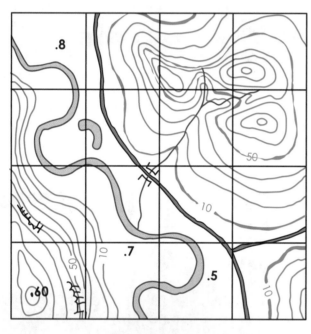

Diagram A
Contour Patterns

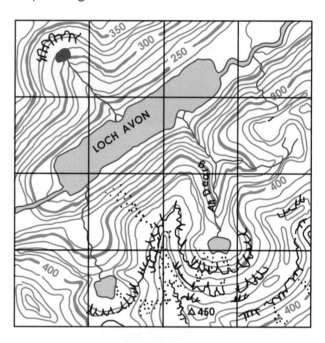

Diagram B
Contour Patterns
Glaciated Landscape

28

Diagram A and B above show a series of contours that shape the landscape. Study the table below to work out what patterns the contours show.

Contour feature	What the contour pattern shows
Closely packed contours	Steep slopes – height increases over a short distance on the ground. Steep gradients
Few or no contours in an area	Flat land, for example along the sides of slow-flowing rivers. Very little gradient
Round circular contours one indide another	A rounded hill.
Long oval contours	Elongated hills.
Large numbers of contours, closely packed	High mountain areas such as the Highlands and uplands of Scotland, or the mountains of England and Wales.
V-shaped patterns	These are found where river valleys cross the map.
Gaps between hills, with contours each side of the gap	These are gaps and passes, or cols and saddles. Roads and paths often go through these where the hills would otherwise form a barrier.

Contours on OS maps are orange in colour. On 1 : 50 000 maps they are set at **10 metres** apart. On 1 : 25 000 maps they are set at **5 metres** apart. The height between contours is called the **contour interval**. The difference between the highest and lowest points identified along a line on the map is called the **range of altitude**.

 CREATING A CONTOUR EFFECT

You can use a potato to create the effect of contours, and this will help you to understand how contours work in the landscape. Cut a large potato in half, lengthways. (As this should be done using a sharp knife, check with an adult before you start cutting.)

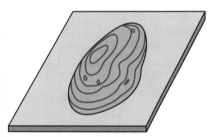

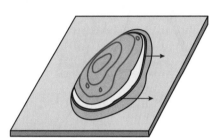

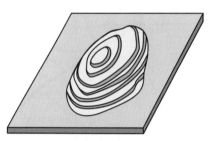

1. Slice the potato

On a chopping board, make thin and thick slices of half a potato, alternating the thick and thin slices.

2. Remove alternate slices

Remove each thin slice to create a stepped effect in the half potato.

3. Paint the slice edges and print

With the thin slices removed you can see how contours are used on a map. You can paint the edges of the potato and print each contour on a piece of paper.

29

SCALE ON THE MAP

Maps are drawn to scale; they are effectively a bird's-eye view of the landscape. If the scale of the O.S. map is 1 : 50 000, for every **one cm** measured on the map there will be **50 000 cm** on the real ground. Maps have a grid and each grid square is one km². 50 000 centimetres is half a km, so the map scale could also be written as 1 cm = 0·5 km. Similarly, every 2 cm measured on the map equals 1 km on the real ground. The grid therefore is a kilometre grid.

 **MAKE THE LINK**

Scale is important in other subjects too. In Technology we use scale when undertaking graphic design projects. We may create scale models or do technical drawings to scale. Scale is used in art for drawing things accurately in proportion.

 Use an OS map (see the Stirling map on page 30) to work out distances as the crow flies (straight line) and by road, rail or on foot. Using the headings in the table below, draw a table and fill it in for eight journeys of your choice.

Place to place	Grid reference	Crow flies distance	Road distance	Rail distance	Foot distance	Map distance
Stirling Castle to Wallace Monument						
...						

USING AN OS MAP

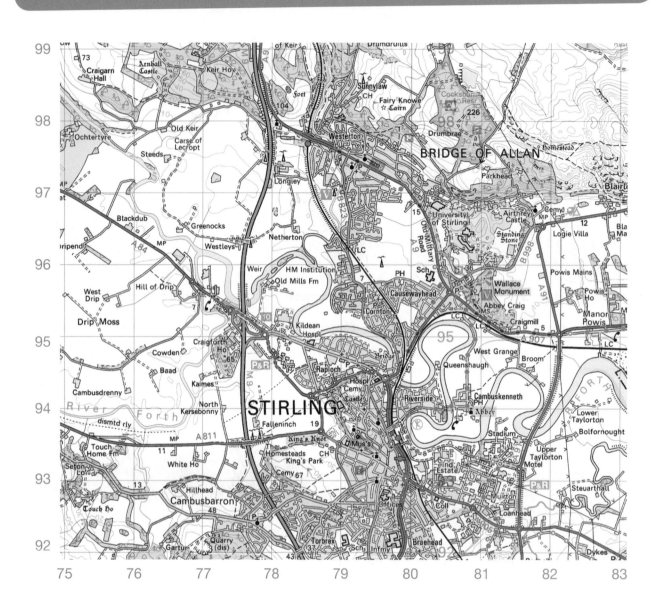

ASSESSMENT B:

Study the map of Stirling above.

1. Describe the relief (shape of the land) in the north-east corner of the map. Hint: mention such things as the steepness of the ground, the shape of the hills, the direction the hills run from and to, the high and low points (using grid references). **(4 marks)**

2. What do you notice about the River Forth in grid squares 8094 and 8095? **(2 marks)**

3. Study the site of Stirling Castle at 792941. Why do you think this location was chosen for the site of a castle? **(2 marks)**

4. Study the site of Stirling University 8096 and 8196. Why do you think this is a good place to build a university? Give three or four developed reasons. **(4 marks)**

TOP TIP

It is important to write developed reasons when answering questions. A developed reason tries to explain why something happens. It requires extended writing and uses key phrases such as, 'this is because ...' or 'a reason for this is ...'

SKETCH MAPS

Drawing a sketch map based on an OS map can be a useful way of extracting information from a map and putting it to another use, for example supporting an essay or an investigation. Follow the steps below to create a simple sketch map of the Stirling area.

1. Draw a grid based on the original map (you may wish to make the grid squares on your map bigger and reduce their number compared with the original map).

2. Use the grid as a guide to position features taken from the map. (This will help you create an accurate map.)

3. Sketch in a few contours to show relief. Try to show groups of contours and try to show low and high ground.

4. Add in any rivers, again using the grid to keep things accurate.

5. Draw in main roads and railways but not all roads.

6. Shade in any town or other major settlement sites.

7. Shade uplands brown and lowlands green to highlight relief features.

8. Label any significant sites and important places. You can develop a map key to hold this and other information.

9. Add any annotations to the key that explain what can be seen on the map.

10. Draw on a scale for the map and a direction indicator. Write a title for the map in an appropriate place.

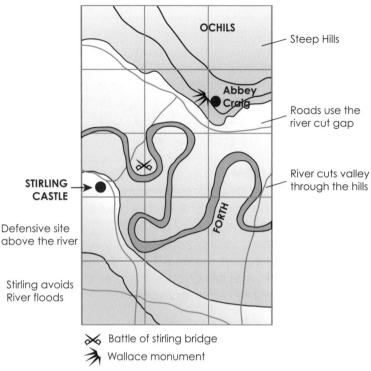

SKETCH MAP OF THE STIRLING AREA

- Steep Hills
- Roads use the river cut gap
- River cuts valley through the hills
- STIRLING CASTLE
- Defensive site above the river
- Stirling avoids River floods

 Battle of stirling bridge
Wallace monument

31

MAKE THE LINK

Statue of Robert the Bruce at Bannockburn

History

- A sketch map allows us to see the landscape where a famous battle in history was fought without all the modern-day map clutter.

- Use this map in History to help your studies of the Scottish Wars of Independence and William Wallace.

INVESTIGATING EARTH FORCES

Learning intentions: In this section you will learn about the structure of the Earth and some basic geology. This will allow you to understand how earthquakes and volcanoes occur around the planet. You will also learn how to put an investigation together where you follow your own interests and research a series of related topics so that you can draw your own conclusions.
Experiences & Outcomes: *Having investigated processes which form and shape landscapes, I can explain their impact on selected landscapes in Scotland, Europe and beyond.* **Soc 3-07a**

BASIC GEOLOGY

The Earth is made up of rocky material but the diagrams below show that much of the structure of the Earth is not made of solid rock. The surface of the planet on land and beneath the oceans is dominated by the rocks of the Earth's crust. There are three basic types of rock: igneous or fire-formed rocks; sedimentary rocks created when other rocks are weathered or eroded by natural forces or as layers of organic material that convert to solid rock; metamorphic rocks. Metamorphic rocks are existing rocks that are changed by being put under great pressure or subjected to great heat in the Earth's violent movements over time.

ASSESSMENT C

1. Copy and complete the table below. **(5 marks)**
2. Find out and write a short paragraph about three of the rock types shown in the table. **(5 marks)**
3. Either:
   (a) Find out about the local rock types in your area. Examine a geology map that shows the rocks. Write a short illustrated account about local geology. **(10 marks)**

   OR

   (b) Collect three rocks from around your school grounds or from your local area. Identify them with the help of your teacher or a local expert (museum visit). Create a Trump card for your rock that includes a digital photo of it complete with labels and descriptions. **(10 marks)**

Rock type	Category	How the rock was formed
Granite	Igneous rock	Deep underneath volcanoes
Limestone		In the ocean from the remains of sea creatures
Sandstone		Deposits of sand by the sea, by the wind and by rivers
Mica schist	Metamorphic rock	Changed by heat and pressure from sedimentary rocks
Basalt		Lava from volcanoes and fissures
Marble		Limestone changed by heat and pressure
Coal		The remains of the bodies of trees compressed over time into rock
Chalk	Sedimentary rock	Made from the bodies of tiny sea creatures
Quartz		A hot volcanic mineral
Gneiss		Changed by heat and pressure from rock-like granite

Use modelling clay to create your own model of the Earth and its structure.

1. With red clay, create the inner core.
2. Roll out a flat sheet of pink clay and carefully wrap it around the core. Roll the whole model into a ball.
3. Repeat this procedure for each of the six layers. Try to estimate the thickness of each layer as you roll out each slab. This will give you a rough idea of the relative thickness of each layer in your finished model.
4. Create continents on your completed sphere; try to do this as accurately as you can. Use a craft tool or blade to cut a quarter-slice out of your model. Try to avoid cutting into the core so that it is left whole. You have now revealed the structure of the Earth.

1 Crust
2 Mohorovic Discontinuity
3 Asthenosphere
4 Mantle
5 Outer Core
6 Inner Core

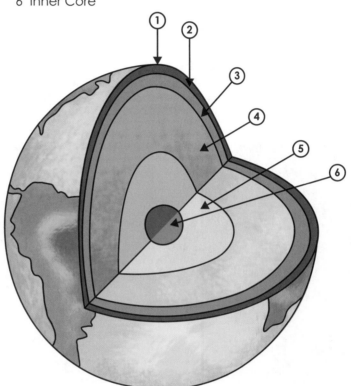

Complex Diagram of Earth Structure

33

INVESTIGATING IN SOCIAL STUDIES

PUTTING YOUR INVESTIGATION TOGETHER

In Social Studies, and Geography in particular, you will be called upon to write **investigations**. To do this well you will need a good understanding of what is required in an investigation. One of the most common investigations teachers set involves finding out about **earthquakes** and **volcanoes**.

THE INVESTIGATION MODEL

There are some basics that you will need to take into account when completing an assignment or homework exercise involving an investigation. This section is based on the Investigation Model given below.

PLANNING
- Work out a general theme
- Decide on a working title
- Lay out 5-10 Study Questions or chapter headings
- Work out a logical order

GATHERING INFORMATION
- Find a range of sources
- Select parts of the sources you will use
- Get different points of view relating to each study question
- Summarise each point of view or take short notes

PRESENTING FINDINGS
- Write out your aims (say what you intend to do in the report)
- Write an introduction
- Include maps diagrams and photos throughout the report
- Write a body of the report
- Write a conclusion
- List your sources

PLANNING

1. Try to develop you own angle on an investigation such as, 'How were the people of Montserrat affected by the eruption of the volcano?' **or** 'What lessons can we take today from the eruption of Mount Vesuvius in Roman times?' Try to create a snappy interesting title.
2. Study questions should involve **trigger words and phrases**, which encourage you to find answers. **Where** (did the event happen)? **Why**? **When**? **What** plates were involved? **What** effects were experienced? **How** were people affected? **Were** rescue efforts made? **Was** aid sent effective in the **short term** and in the **long term**? **Will** the event be repeated in the future?

GATHERING INFORMATION

1. Use a number of **different sources** of **different types**, e.g. textbooks, newspapers, magazines, videos and DVDs, computer programmes, the internet, your teacher, a letter requesting information. This is the key to an interesting and original investigation.
2. Use highlighters or underlining to identify **key sections.**
3. Take short notes from audio-visual sources, gather images for later use. Never **cut and paste** or **copy directly** from sources as this is cheating. Instead, read sources and learn what they tell you. Write out what you learned in your own words in note form, preferably taking from two or more sources to create something new.
4. **Add value** to maps, diagrams and photographs by annotating them to explain what they show or add extra information from other sources.

34

PRESENTING YOUR FINDINGS

1. Work out the best order for your presentation. Remember that you are telling a story: get things in the right order so that the story fits together.
2. Decide your style of presentation:
   - a time-based narrative;
   - point and counterpoint (e.g. 'Smith said this in his newspaper article but Waugh states that ... and the photo (picture A) supports this point of view.');
   - diagrammatic and pictorial.
3. Set out clear aims for what you intend to do in the presentation. Lay out your study themes or questions. Write an introduction that tells the reader, for example, where your study is based and why you have chosen this topic.
4. Write up the investigation in a logical order, perhaps around a time line or in the order of your study questions. Be sure to refer to every diagram or photo you have used to support what you say.
5. Bring your work to a close with a conclusion that summarises what you have said or that looks at the future and suggests what might happen next. Alternatively come to a judgement or series of judgements based on what you found out.
6. Set up a list of sources.

LIST YOUR SOURCES

Francis, P. (1995) *Volcanoes: a Planetary Perspective*, Oxford University Press, Oxford.
Savage Earth, Channel 4.
Smith, Montserrat. *National Geographical Magazine*, June 1980, pp. 207–11.
http://www.mvo.ms
http://volcano.oregonstate.edu

MAKE THE LINK

• You could find a lot out about life in Roman times to support your studies in History by studying the eruption of Vesuvius in AD 79 and the destruction of the Roman towns of Pompeii and Herculaneum. Similarly, you could support you studies in French by studying Mont Pelée and the 1902 eruption. Volcanoes, their chemistry and physics, and their effects on global climate may relate to work done in Science. You may wish to look at the way geothermal power can be used as a source of alternative or renewable energy by studying geothermal power plants.

DID YOU KNOW?

Volcanoes and earthquakes occur next to tectonic plates. It is a good start in an investigation to find out what plates are involved and in what ways the plates are moving near your case study.

OUR EVERYDAY LIVES

Evidence of Earth forces activity is all around us. Scotland has many examples of volcanic hills, such as Arthur's Seat in Edinburgh, Stirling Castle rock, the Ochil and Sidlaw Hills or the Isle of Skye. We often visit volcanoes while on holiday, for example Mount Teide in Tenerife or Mount Vesuvius in Italy. We may even find ourselves in an earthquake zone, for example in Turkey, Italy or California. Many TV programmes are made about volcanoes and eruptions, and earthquakes often feature in the news, especially if they involve loss of life and dramatic events such as tsunamis.

DEVELOPING AN EARTH FORCES INVESTIGATION

MONTSERRAT

This section uses a case study of the island of Montserrat in the Caribbean (see right) to illustrate some of the elements that might be used in an investigation of a natural disaster.

In July 1995, the Soufrière Hills Volcano became active after a long period of dormancy (see below). This created an emergency in which two-thirds of the people of the island were forced to leave their homes. Many of these people were unable to return to the island as the volcano continues to erupt and continues to cause damage.

36

The key to this investigation is a series of questions about the volcano that help to explain why it erupted and what effects the eruption has had. In many ways this investigation outline could form a template for you to follow in your own investigation of a natural disaster involving earthquakes or volcanoes.

INVESTIGATION PLANNING MATRIX

Start by asking a question that gives a background to the investigation such as, 'Why did the Soufrière Hills volcano erupt and what were the effects of the eruption on the landscape and people of Montserrat?' Use the hints in the **Investigation Planning Matrix** below to guide you through the process. This will give you help and ideas but will allow you to develop your own strategy for any investigation, not just one involving earth forces.

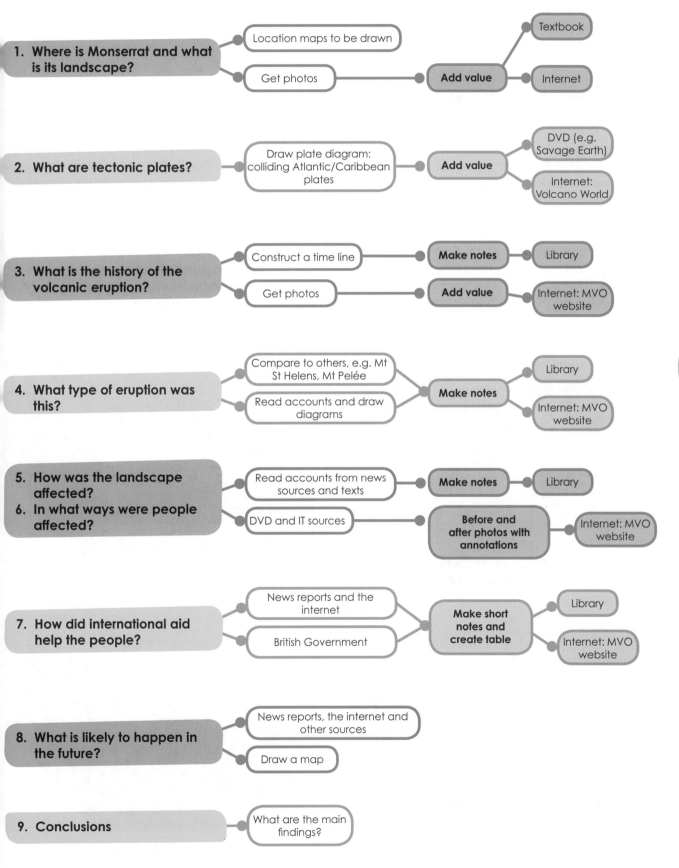

OUR DYNAMIC EARTH

WHAT IS THE EARTH'S STRUCTURE LIKE?

The model of the Earth's structure on page 33 shows four main layers, the solid **crust**, the plastic **mantle** and the liquid **outer core** and solid **inner core**. The interior of the earth is a hot-bed of radioactivity and is under enormous pressures, which makes it extremely hot.

HOW DO PLATE TECTONICS WORK?

Studies of earthquakes and volcanic activity revealed that the Earth's crust is broken into large sections of crust called **crustal plates**. These come together like the pieces of a jigsaw but they are constantly moving and interacting with each other. The edges of the plates are called **plate margins** and it is at these points that most volcanoes and earthquakes occur. Plates beneath the ocean are called **oceanic plates** and are thin. Plates that support the continents are called **continental plates** and are much thicker.

The mantle below the plates is in slow but continuous movement due to convection currents set up in the mantle by the heat of the Earth's interior. The crustal plates float on the mantle so as the mantle moves, so do the plates. The plate diagrams below show the main ways the plates move and interact with each other. The plate diagrams and the map showing plate boundaries are good starting points for studying earthquake and volcanic activity, and are useful in supporting investigations.

38

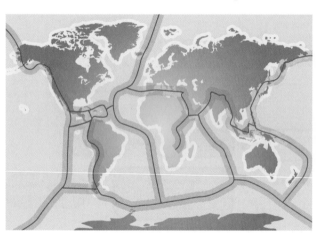

Plate boundaries

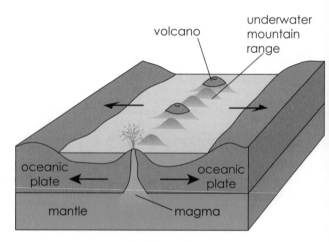

Parting plates

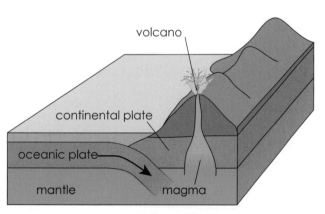

Colliding plates

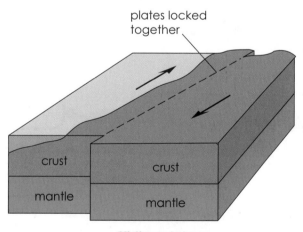

Sliding plates

CO-OPERATIVE LEARNING: 'JIGSAW EARTH'

Learning intentions: in this section you will learn about volcanoes and how they erupt. You will be able to create models of volcanoes using waste materials.

Social skills: You will work with others to complete parts of a task so that if one succeeds, then the group succeeds.

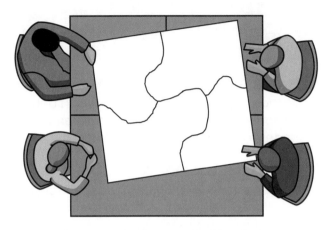

In a 'jigsaw' activity a task is divided up so that each member of a group is given part of the task to perform. Once each person has completed their individual section (like a piece of a jigsaw) the group comes together and shares what each person has learned or found out. In this way the jigsaw is completed and everybody has learned each of the parts.

Allocate one of the parts to each member of your group:

• **Parting plates** – explain what happens and what effects they have in terms of volcanoes and earthquakes. Find one example of this type of plate movement.

• **Colliding plates** – explain what happens and what effects they have in terms of volcanoes and earthquakes. Find one example of this type of plate movement.

• **Sliding plates** – explain what happens and what effects they have in terms of volcanoes and earthquakes. Find one example of this type of plate movement.

• **Trace and complete** the jigsaw below – explain what evidence can be found to show that the continents involved were once joined together.

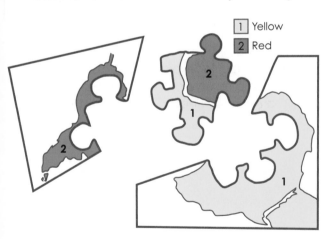

1 Yellow
2 Red

Come back together and **share** your findings. Make notes and draw sketches to help you record what your group have found out.

Study the diagram on page 45 of the Japanese earthquake of 2011. As a **group**, discuss, describe and explain what happened to cause the great earthquake and tsunami in this part of the world.

39

VOLCANOES: LET'S GET VERY ACTIVE

WHAT IS THE STRUCTURE OF A VOLCANO LIKE?

Models of volcanoes, complete with labels and annotations showing you what goes on inside, can be found in textbooks and on the internet. A much more active and impressive way to show your understanding of volcanoes is to build a model of one.

HOW TO BUILD A MODEL VOLCANO

1. Copy the outline of the volcano model diagram (see below) onto A4 paper or thin card. Alternatively, take a photocopy and enlarge it to a size that suits you. Cut around the outline.
2. Find a reference diagram in a textbook or on the internet that shows you a cross section of a volcano (web search: volcano cross section).
3. Use the reference diagram and photos of volcanoes to colour your model. You should use lots of red, orange, yellow, grey and black to show magma, lava, ash and volcanic rock. Try to make your colours as vivid as possible.
4. Fold the flap on the side of your painted or coloured model and glue the top surface. Next stick the flap behind the left-hand slope of the volcano to complete the cone. Add labels and annotations to your model and write a piece about it, explaining how the volcano formed and what parts of it do.
5. Add special effects:
   • Smoke effects – using painted cotton wool.
   • Ash and rock – glue sand and small stones onto the surface of the cone.
   • Lava flows – run PVA glue mixed with red paint down the cone.
   • Cut the crater wide enough to insert a party popper for spectacular explosive effects.
   • Attach a small open container beneath the crater to create a reservoir; add a little bicarbonate of soda plus some red food colouring, then add a small amount of vinegar and watch the volcano erupt.

CO-OPERATIVE WORKING: VOLCANO MODEL ACTIVITY

Social goals: Working as a team; coming to an agreement; dividing labour.
Form a group of **four**. Agree a team leader, a resource manager, an IT manager, a space manager. Read through the instructions and break the task up into sections and allocate work. The team leader should solve any disagreements and the IT manager should do any computer research.

40

MAKING A BOX MODEL

You will need:
- an A4 paper box and its lid
- scissors
- a newspaper
- PVA glue (watered down to a thin consistency) or paste
- water paints in volcanic colours

Cutting the cardboard

1. Lay the box lid to one side to act as a base for the model.
2. Carefully break the A4 paper box apart at its glued joints to preserve the cardboard panels.
3. Draw a volcano cross section that stretches the length of the box lid on one of the panels. Use a reference diagram to help, or base it on the diagram on page 40. Be sure to add two or three tabs below the cross section as these will stick through the box lid later and hold your model in place.
4. Cut the cross section out of the cardboard panel.
5. Use the cross section as a template and trace further cross sections onto the remaining cardboard panels. Make as many cross sections as you can out of the cardboard.
6. Keeping one cross section whole, cut all the others up the middle to create a series of half sections that will be fixed to the lid by their tabs to make up the volcanic cone.

Building the frame of the cone

1. Use the whole cross section and align it along the front edge of the box lid. Next mark where the tabs will slot into the box top with a pencil then cut slots in the box lid to take the cardboard tabs on the bottom of the section. Slot the cross section in place.
2. Repeat this process with the half sections, lining them up with the centre of the upstanding whole cross section. Assemble the frame of the volcano by slotting all the half sections into place.

Clothing the volcano

1. Cut some of the newspaper into thin strips about 1 cm wide. (Lay out a piece of newspaper as things are going to get messy!) Coat one of the strips of newspaper with watered-down PVA glue and place it on the volcano framework. Repeat this procedure until the framework of sections is covered and there are no gaps visible around the cone. Make sure you have created a good-sized crater in the top of the model for later.
2. Tear off small pieces of newspaper and stick these onto the cone to create a smooth surface. Set the volcano aside to dry for 24 hours.

Finishing the model

1. Use paints and a series of reference diagrams (from the internet or from a textbook) to colour the model. Add visual effects to make the model come to life.
2. Create a visual display by writing about the volcano you have created and explaining what is going on. Add labels to identify the volcano's various parts.

EARTH FORCES: EARTHQUAKES

Learning intentions: in this section you will learn about earthquakes – both why they occur and how they can have devastating effects on people.
Social skill: here you will be able to work on coming to agreement with others by working collaboratively.

WHAT HAPPENS IN AN EARTHQUAKE?

Two recent earthquakes – the Haiti Earthquake of January 2010 and the March 2011 Japanese earthquake and tsunami – reveal the damage that earthquakes can do. They are much more dangerous than volcanic eruptions because there is no warning given and people have little chance to escape. Volcanic eruptions are usually preceded by a period of increasing volcanic activity. With earthquakes there is not only a lack of warning but the event is also followed by a series of aftershocks, which add to the damage and make rescue operations more difficult and dangerous. Earthquakes can create tsunamis, which wash over low-lying coastal areas and can potentially kill or make many people homeless hours after the initial earthquake.

Earthquakes occur mainly close to plate boundaries. Tension builds along the fault lines found at the plate edges where two or more plates move against each other. Tension and pressure build up until something has to give and the plates move and reposition themselves. Shockwaves move away from the epicentre of the earthquake deep below ground and travel away from it and across the Earth's surface. Plates can move by sliding past each other along a fault line, such as the San Andreas fault in California.

San Andreas fault in California

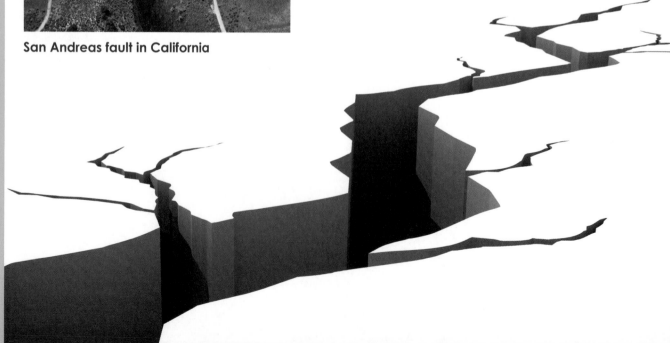

42

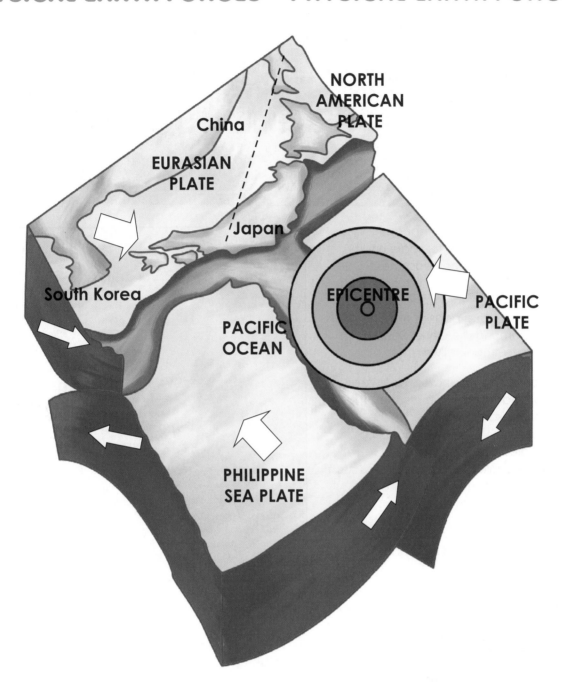

The Great Japanese Earthquake 2011

Other plates move over and under each other, like the Atlantic Plate and the Caribbean Plate or the plates involved in the 2011 Japanese tsunami. Plates moving beneath the ocean create tsunamis, where the movement of the plate displaces and disturbs the water column above it, creating waves that drive towards the shore of nearby and even distant countries.

43

CASE STUDY: THE JAPANESE EARTHQUAKE AND TSUNAMI, MARCH 2011

In March 2011 a massive earthquake measuring 8·9 on the Richter scale hit the east coast of the largest Japanese island, Honshu. The earthquake triggered a 10-metre-high tsunami that rushed several kilometres inland along parts of the coast, with the area around the city of Sendai being most catastrophically hit.

A month later, it was still unclear as to how many people had died in the disaster: the Red Cross organisation (involved in the disaster relief effort) estimated that 8000 people had died and 12 000 people were still missing. The scale of the disaster was such that it would take months before the true death toll would be known.

In 1923 the Great Kanto Earthquake killed 140 000 people in this area of Japan.

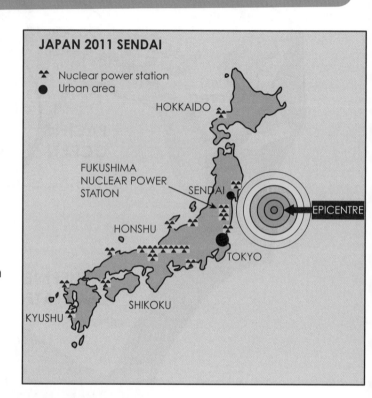

JAPAN 2011 SENDAI
- ☢ Nuclear power station
- ● Urban area

HOKKAIDO

FUKUSHIMA NUCLEAR POWER STATION

SENDAI

HONSHU

EPICENTRE

TOKYO

SHIKOKU

KYUSHU

44

CO-OPERATIVE LEARNING: A GRAFFITI BOARD ACTIVITY

Social skill: To arrive at agreements with others.

Learning intention: To devise a series of suitable investigative questions about the great Japanese earthquake of 2011.

Resources: large sheet of paper on the tabletop (**graffiti board**), marker pens, sheets of A4 paper or jotters. The group task is to devise a series of questions about the 2011 Japanese earthquake. Form a co-operative table group of four. Each group member should write out up to **10 research questions** about the earthquake and its aftermath. Share your study questions with your shoulder partner, taking turns to tell your partner one question at a time. On a **separate piece of paper**, write down the questions suggested by the pair, combining any questions that are the same or similar. Agree between you the **10 best questions** and mark these down on your writing paper. Now share your questions, taking turns with the other pair in your group. Come to agreement over what would be the best questions for an investigation of the earthquake disaster. Write these out on the **graffiti board** sheet. Justify why you think each question is a really good one for an investigation.

Share your chosen questions with the rest of the class, justifying your choice of question. Enter the class questions on a **dotmocracy** grid, again combining any that are similar or the same. Use dotmocracy to choose a group of the most popular and best questions for the class to use in an investigation.

THE TSUNAMI

The scale of the Sendai earthquake is hard to really appreciate. It was the sixth-biggest quake since the year 1900. It generated a 10-metre-high tsunami that washed over the northern Sendai coast. Beyond the immediate death toll, it destroyed whole districts of coastal towns, swept away thousands of homes, displacing 215 000 people to emergency shelters, damaged farmland and growing crops, created serious fires and caused catastrophic damage to four of the six Fukushima nuclear reactors. See the diagram on page 44.

The earthquake threw up a number of startling facts and statistics:

- It took over three weeks to get to grips with the nuclear emergency at Fukushima. This was achieved by restoring electric power to cooling pumps.
- 14 000 houses were destroyed, 100 000 were damaged and 340 000 people were evacuated from their homes.
- 256 819 houses were left without electricity.
- Over 1 million people were left without running water.
- Around 50 000 rescuers were deployed to the area within days of the earthquake.

The earthquake and tsunami caused massive levels of destruction

45

ASSESSMENT E

Create a graphic display poster that explains the causes and effects of the Sendai earthquake. Start by searching the internet to find photos and graphics, such as maps and diagrams. Avoid 'cut and paste' activities: instead use the graphics you find to create your own versions, by reconstructing and combining diagrams and maps using programs like 'Paint'. Add value by taking elements of one source and adding it to another. In addition, add even more value by providing your own annotations and labels to your photos and diagrams. Remember to identify and acknowledge your sources on the page.

Before starting your poster design, create a sketch plan with roughs of the graphics and annotations you are intending to use. Discuss your draft with another person or with your teacher and use any advice you get to help you complete the task. **(20 marks)**

MAKE THE LINK

In Modern Studies we study how international aid can be sent to help in a disaster.

In Economics, the impact of an earthquake or other disaster can be analysed in terms of the effect on the economy.

DID YOU KNOW?

Some British schools have been given seismographs by the British Geological Survey and these picked up wave traces from the earthquake. Tsunamis travel over vast stretches of the world's oceans. For example the Krakatoa tsunami of 1883 washed up on the shores of Britain as a series of larger than normal waves some days later.

CASE STUDY: THE HAITI EARTHQUAKE

Read the following newspaper report.

Haiti's Earthquake Crisis

On 12th January 2010, at 16:53 local time, an earthquake measuring 7 on the Richter scale hit the Caribbean island of Haiti. the epicentre was near the town of Leogane, 15 km to the west of the capital city Port-au-Prince. As many as 250 000 people were killed by the earthquake and its immediate aftermath. Over 1 million people were made homeless and hospitals were overwhelmed by more than 300 000 injured victims. The proximity of the earthquake to a major city made the disaster all the greater. Cheaply constructed concrete buildings that had dominated the city centre simply crumpled with the shocks.

Haiti is one of the poorest nations on earth and was unprepared for such an event. Rescue facilities were scant and it took days and weeks for international aid to arrive, further increasing the death toll. People quickly found themselves without food or water as supplies were disrupted. The centre of government was hit, so relief efforts were hard to coordinate.

Port-au-Prince, Haiti's capital city

HOW DOES INTERNATIONAL AID HELP IN A DISASTER?

The Disasters Emergency Committee (DEC) is a charity organisation that co-ordinates the efforts of a number of agencies and charities involved in helping after disasters have occurred.

> 'The DEC has raised £101m since the earthquake on 12 January. Only the 2004 tsunami prompted a bigger response from the British public. So far, £30m has been spent on providing emergency assistance to 1.2m people. The DEC's Brendan Gormley said "a long and painful journey" still lay ahead.'
> (Source: http://www.bbc.co.uk/news)

After a disaster like the 2004 Boxing Day tsunami in the Indian Ocean or the earthquake in Haiti strikes, help begins to arrive. The United Nations organisation uses UNDRO, the UN Disaster Relief Organisation. Governments send shipments of food water, medicines, tents and equipment to help. International agencies, such as the International Red Cross, send in medical teams and rescue specialists. Soldiers, police and rescue services within the country are usually the first on the scene with heavy equipment and helicopters to give access to the injured and to get in necessary supplies.

Many people in Britain donate to the DEC or to other charities, often responding to appeals in the media. It is often ordinary people like you and me who make the biggest difference to people hit by such emergencies when we donate money to relief campaigns.

It can take many years for a very poor nation like Haiti to recover from a natural disaster. Countries can be so disrupted that it has long-term effects on health and development. The earthquake in Haiti was followed later in the year by an outbreak of the disease cholera because of poor sanitation and contaminated water supplies.

ASSESSMENT F (25 MARKS)

1. Undertake an investigation into a major earthquake event of the past such as the Great Japanese Tsunami.
2. Study news reports on the internet that reveal how disasters are covered in the news. Look at how diagrams and photos are used to help explain things. Write your own account of the Haiti disaster or investigate another disaster as a journalist and write a newspaper article.
3. Write a two-page section for this book but make it a section you would like to see included relating to disasters. Send your 'copy' to the publishers by e-mail (www.leckieandleckie.co.uk).
4. Follow a breaking news story of an earthquake or volcano eruption. Collect newspaper and TV/radio reports and write summaries. Create a **web page** to present the news about the unfolding disaster. Get your co-operative group, class or school to raise money for disaster relief and track how it is spent. In this way you will raise awareness and show you are a global citizen.

QUESTIONS TO THINK ABOUT AND RESEARCH:

1. Why do different sources offer different casualty figures for an earthquake?
2. Why does information improve as time goes on when dealing with natural disasters?
3. Why is it important to get aid to the scene as quickly as possible?
4. How are people killed and injured in earthquakes?
5. Why are there no serious earthquakes in Britain?
6. Why is securing water supply one of the first priorities in a disaster zone?
7. How does the Richter scale work?

Seismographs record earthquakes and tremors

MAKE THE LINK

- In English we use functional writing, and learn to write in other styles such as those used by the print media. It is important to check your information by verifying your sources; in other words making sure things you write are true by checking the facts against other sources.

- In Modern Studies, international development is a major topic of study along with the politics of trade and aid.

- Media Studies looks at the coverage of significant news events.

- How photography is used clearly relates to subjects like Art and Graphic Communications.

- Geology has close links with the Sciences and will crop up in these courses in various ways.

DID YOU KNOW?

47

The web resources on earthquakes are amazing. If you search carefully you will come up with some of the best educational sites on the internet. The reason for this is that children and young people all over the world study these amazing and terrifying features. No more so than in the United States, where the subject is part of Earth Science. Check out this web link for example: http://earthquake.usgs.gov/earthquakes/world/world_deaths.php. This is the website of the US Geological Survey, who provide a wide range of geographic information for schools.

OUR EVERYDAY LIVES

Geology and Earth Science can be studied at many universities in Scotland and the UK. People studying geology can go on to work in the oil industry or industries involved with mining and resource exploitation. Large civil engineering projects often employ geologists. Geologists also work on monitoring volcanoes and developing hazard maps. Like geographers, they use GIS as an effective tool in their work. In the future geologists and earth scientists will be key players in space exploration.

GLACIERS AND MOUNTAIN LANDSCAPES

Learning intentions: in this unit you will discover how erosion shapes the surface of the Earth and how this is effectively done by glaciers.

Loch Avon in the Cairngorms

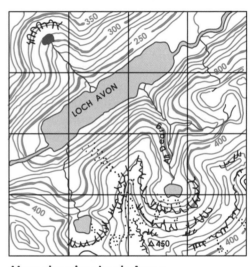

Map showing Loch Avon

48

The United Kingdom and most of Northern Europe has spent much of the last two million years in an ice age, and geographers think that although the ice melted 10 000 years ago it could return in several thousand years.

Ice acts like a giant grinding machine, tearing the land surface apart and wearing it down in a process called **erosion**. After an ice age ends and the glaciers have gone the landscape is completely changed.

Loch Etchachan in the Cairngorms

The mountains are the best places to see the effects of the ice age and the grinding glaciers. In Scotland, the Highlands bear the scars of the erosion that took place thousands of years ago. The most spectacular features formed are **u-shaped valleys** and **corries**. All over the world, similar features were left behind by the glaciers and ice sheets of the ice age. If you visit the Alps in Europe or the Rockies in Canada and the USA you will see the same glacial scenery. Wherever high mountains are found the evidence of the Ice Age is found as well.

HOW DO GLACIERS CHANGE THE LANDSCAPE?

Before the Ice Age, the landscape of Scotland was created by **river erosion**. As the climate got colder, ice began to collect in the heart of the mountains in areas like Rannoch in the Grampian Highlands. Small glaciers grew in cold north-facing hollows on the hillsides, because as the climate chilled down snow falling in the winter did not melt and instead built up in layers, changing into **glacier ice**.

Glacier ice moves down slopes under gravity, forming a slow-moving river of ice. The glaciers occupied the existing v-shaped river valleys and stripped away any soil and loose material, then began to drag across the solid rock below. This led to a process called **plucking**, where large chunks of rock are torn away from the landscape by moving ice. You can see some of this material in the photos of Loch Avon and Loch Etchachan on page 48. Look for the big boulders lying on the hillsides.

All the rocks, stones sand and boulders mixed up in the ice act like a giant grinding machine and this wears away the land in a process called **abrasion**. Over thousands of years of glaciation, the valleys change their shape into much bigger u-shaped valleys with steep sides and flat floors. These form some of our most spectacular scenery.

THE FORMATION OF GLACIATED LANDSCAPES

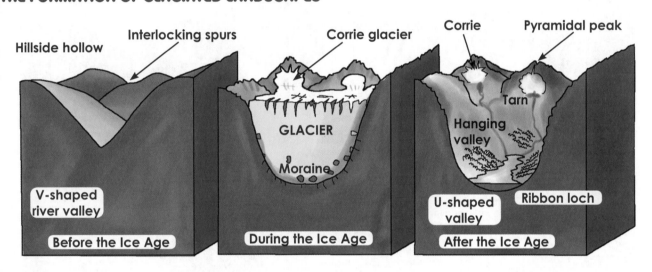

49

DID YOU KNOW?

The first national park in the US was Yellowstone, established in 1872. It was Scotsman John Muir who convinced President Theodore Roosevelt to create Yosemite national park. Muir is regarded as the father of modern conservation. He came from Dunbar, where he is commemorated in the John Muir Country Park, the John Muir Trust and John Muir Awards.

GLACIERS AND MOUNTAIN LANDSCAPES

HOW ARE MOUNTAIN LANDSCAPES USED?

Britain's mountain areas cover a large area of the country. They form important working environments for many people. They are a breathing space for others who like to visit them from the towns and cities. For other people they are a playground and holiday destination. Energy is generated in them and water supplies for towns and cities are collected and stored there. Upland Britain is also home to our national parks, and here we find important natural landscapes, forests, moors and wildlife habitats.

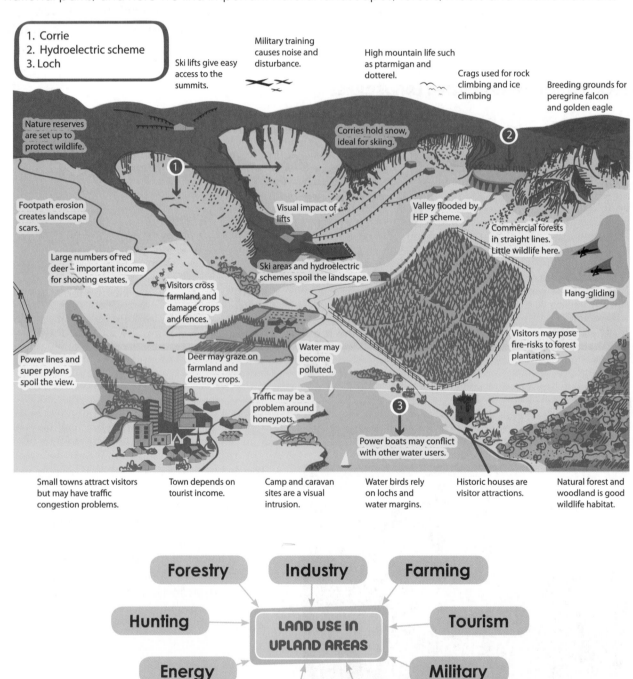

1. Corrie
2. Hydroelectric scheme
3. Loch

Ski lifts give easy access to the summits.

Military training causes noise and disturbance.

High mountain life such as ptarmigan and dotterel.

Crags used for rock climbing and ice climbing

Breeding grounds for peregrine falcon and golden eagle

Nature reserves are set up to protect wildlife.

Corries hold snow, ideal for skiing.

Footpath erosion creates landscape scars.

Visual impact of lifts

Valley flooded by HEP scheme.

Commercial forests in straight lines. Little wildlife here.

Large numbers of red deer – important income for shooting estates.

Ski areas and hydroelectric schemes spoil the landscape.

Hang-gliding

Visitors cross farmland and damage crops and fences.

Power lines and super pylons spoil the view.

Deer may graze on farmland and destroy crops.

Water may become polluted.

Visitors may pose fire-risks to forest plantations.

Traffic may be a problem around honeypots.

Power boats may conflict with other water users.

Small towns attract visitors but may have traffic congestion problems.

Town depends on tourist income.

Camp and caravan sites are a visual intrusion.

Water birds rely on lochs and water margins.

Historic houses are visitor attractions.

Natural forest and woodland is good wildlife habitat.

Forestry   Industry   Farming

Hunting

LAND USE IN UPLAND AREAS

Tourism

Energy

Military

Recreation and leisure   Water storage and supply

50

DEFINITIONS

USING AN EDUCATIONAL WEBSITE

Match the **glaciation key words** to what they mean.

Glacier	An armchair shaped hollow found high on the sides of mountains
Pyramidal peak	A 'river' of ice
	How glaciers erode the landscape
River valley	A deep, steep-sided u-shaped valley with a flat floor cut by a glacier
Corrie	V-shaped valleys with interlocking spurs cut by river erosion
Ribbon lake	
Glaciated valley	Sharp, pointed peaks found in the mountain areas of the world shaped by glaciers
Plucking and abrasion	Long, narrow lakes found at the bottom of glaciated valleys

By using your computer and the internet you can find many excellent educational websites. One example is the Cairngorms National Park Authority website www.cairngorms.co.uk, which has sections especially devoted to the needs of Scottish school students. Find out about national parks and their important place in looking after the environment of Scotland and the UK. The Cairngorms National Park is the largest in the UK.

Use the Highland land-use model on page 50 and the Cairngorms website to find out how uplands in Britain are used by people for work and leisure.

Make comparisons with European national parks, for example the Camargue in France (www.parc-camargue.fr) or search for US national parks such as Yellowstone or Yosemite (www.nps.gov/index.htm)

51

ASSESSMENT F

1. Study the diagram below showing how a glaciated valley forms.
   Match the statements below to the letters on the reference diagram. **(2 marks)**

A steep sided u-shaped valley left after the ice melts	
V-shaped river valley	
Glacier erodes the valley using plucking and abrasion	

2. Explain how one of the u-shaped valleys was formed. You may use a diagram(s) to illustrate your answer. **(3 marks)**

3. Study the Highland land use model above. Choose one of the land uses shown and say why the upland areas are suitable for this activity. **(3 marks)**

A
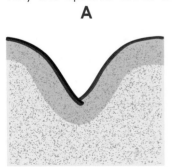

B
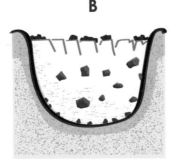

C
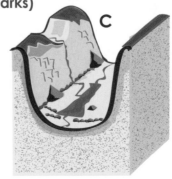

THE WORK OF RIVERS

Learning intentions: in this section you will learn how rivers change the landscape through erosion and deposition. You will be able to identify a wide range of river features and explain how they formed.

THE WATER CYCLE

All the water on the planet has been here since the Earth first formed. In any glass of water the millions of molecules present have circulated again and again: a living example of recycling on a world scale.

HOW DOES THE WATER CYCLE WORK?

The water cycle starts in the seas, where water is evaporated by the sun and is lifted up as invisible water vapour within the air by warm air currents. Air travels on the wind towards the land. As the air travels over land it is forced to rise, especially where there are mountains and hills to push it upwards. Air rising in this way cools and it is now unable to carry as much invisible water vapour as it gets colder. The water vapour condenses and forms clouds, which contain billions of tiny water droplets so small that they float on the air.

Still rising over the mountains, the air continues to chill and the water vapour forms precipitation (snow, rain, hail). This sees the water now returning to the Earth's surface, where it collects into rivers and streams that flow downhill, beginning their return journey to the sea.

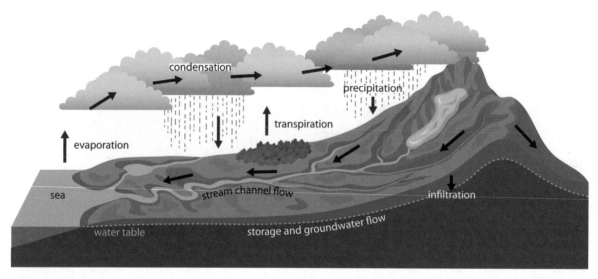

Not all the water goes directly back to its ocean source. Some gets taken up by trees and plants. Some water is stored in glaciers and lakes. A large portion of the flow goes underground and slowly seeps through the soil or finds its way deep underground into permeable rocks, such as sandstone, limestone and chalk. Before reaching the sea, water may linger in swamps and marshes but eventually it will return to the sea – maybe through a river estuary like the Firth of Tay or through a delta such as the River Nile Delta in Egypt.

BUILDING A WATER CYCLE MODEL

You will need:
- two cardboard A4 paper boxes
- PVA glue or paste glue brush
- newspapers
- water-based paints and brushes
- sharp scissors

INSTRUCTIONS

1. Carefully break **one** paper box into its rectangular cardboard panels and set these aside.
2. Draw the sea to mountain-top profile on the two long sides of the box in pencil and cut it out.
3. Cut across the newly created **low end** of the box to join the sea end profiles.
4. Use the profile on the box side to draw in pencil and to cut two or three matching profiles in the cardboard sections you already set aside.
5. Glue these in place. (You may need to use paper strips to support them by sticking them to the profile and to the box sides.)
6. Cut strips of newspaper about 2cm wide and brush these generously with watered-down PVA glue or paste. Lay these strips across the cardboard profile ribs to create the landscape.
7. Finish the landscape by gluing on another layer of newspaper as small patches.
8. Leave the model to dry before painting on a landscape that includes all the elements of the water cycle.
9. Add to the model by putting on details such as forests made of little pieces of sponge or lichens. Glue sand and small stones onto the surface to give realistic effects.
10. Label and annotate your model by creating labels on a computer. Stick these onto the model in appropriate places. Add a description of how the water cycle works onto the side of the box in small type.

DID YOU KNOW?

It's funny to think that the water we drink today was once quenching the thirst of dinosaurs, making up ancient seas and lakes and – who knows – some of the water molecules could have been drunk by a famous person from history such as William Wallace or Mary Queen of Scots.

53

RIVERS AND EROSION

HOW DO RIVERS WEAR THE LAND AWAY?

Rivers wear away the land over time by a process called erosion. Valleys are created by rivers using the rocks, stones and sediments that they carry as cutting tools (**corrasion**), plus the force of the water itself (**hydraulic action**).

River valleys are deeply cut in a v-shape in the mountains and hills but get wider as they move down into the plains. As the river moves from its source to its mouth at the sea there are a whole series of river landscape features that can be identified.

Rivers can be divided into three parts: the upper course, the middle course and the lower course. At each stage of the river the landscape tends to be different.

HOW DO WATERFALLS AND GORGES FORM?

Waterfalls and gorges are found where a river erodes hard and soft rocks at different rates. Hard rock is difficult to erode but soft rock erodes more easily.

The Niagara diagram below illustrates how many waterfalls form where soft and hard layers of rock are involved. The Niagara River, flowing from Lake Erie in North America, drops over the Niagara Escarpment (a cliff). Layers of hard limestone making up the river bed are less easily eroded than the softer shales beneath. The plunging river erodes the softer shales, undercutting the limestones. Eventually the erosive

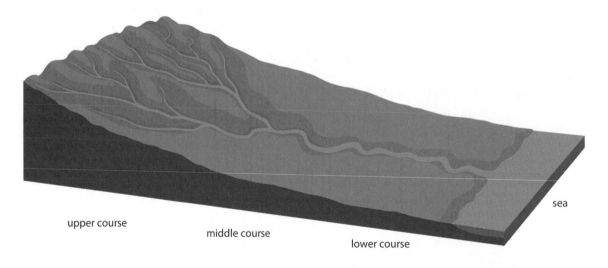

upper course

middle course

lower course

sea

Upper course	Middle course	Lower course
Steep-sided v-shaped valleys with interlocking spurs where the river twists and turns	The river has a narrow floodplain with river cliffs at the side (bluffs)	The river is wider and appears to be slower but has its greatest force now
Gorges are formed as the river erodes the landscape, with rapids and pot holes cut into the bed of the river	There may be meanders that wind from side to side across the valley, helping to erode the landscape	There is a flat floodplain either side of the river. The river will flood over this so embankments may be built to prevent flooding
Waterfalls may form where the right conditions exist	The river both erodes and deposits sediment on the landscape	Meanders cut into themselves to form ox-bow lakes

power of the river undercuts the harder rocks so much that it can no longer support itself and collapses into the plunge pool below. Whirling rocks in the plunge pool undercut the base of the waterfall and cause more collapse, so the waterfall steadily retreats upstream towards Lake Erie, creating the long steep-sided Niagara Gorge below the falls.

Niagara Falls

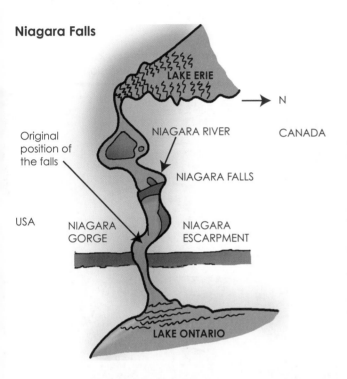

55

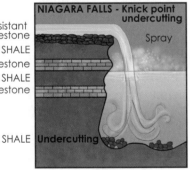

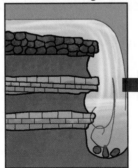

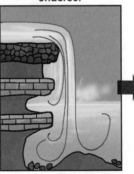

MAKE THE LINK

• In Science we study renewable energy and electricity. In both Geography and Science we are concerned about the environmental impact of power generation. Generating hydro power is environmentally friendly in some ways, but many people argue that it has an effect on the river ecosystem downstream. The reduced flow over Niagara Falls illustrates the down side of power generation as the waterfall has lost most of its natural power.

DID YOU KNOW?

Only about a third of the actual flow of the Niagara River goes over the Niagara Falls. Most of the river's flow is diverted through hydro electric tunnels to provide valuable renewable power for large areas of Canada and the United States. The falls are still spectacular but are not as big in terms of water volume as they might otherwise be.

INVESTIGATING WEATHER AND CLIMATE *

INTERPRETING WEATHER MAPS AND CHARTS

The weather map opposite is a synoptic chart. Note its date, as this already tells you what weather to expect. January is a winter month so relatively cold conditions would be expected. The lines on the map are isobars, which are lines joining places with equal atmospheric pressure. They are the weather map equivalent of contours, and show high and low points in terms of pressure on the map. Large circular areas are pressure systems, which will either be high or low pressure systems (depressions). On the map, Britain lies under a high pressure system, which is centred to the north-east of Shetland. A low pressure system is approaching Britain from the west and lies off the coast of Ireland out in the Atlantic. This pressure system has two fronts associated with it: firstly a warm front, followed by a cold front. These low pressure systems are in the westerly air flow that normally affects Britain so they move from west to east across the map.

MAKE THE LINK

Weather study (Meteorology) is also part of physics. Weather-recording instruments were all invented by physicists who are perfectly at home dealing with atmospheric pressure and other weather phenomena.

HIGH PRESSURE SYSTEMS

High pressure systems (anticyclones) bring low wind speeds as the isobars are usually far apart. In winter they bring cold, dry and frosty conditions, especially if the skies are cloudless at night. In summer they are associated with hot dry days, again with little wind, but you may see thunderstorms if the conditions are right.

LOW PRESSURE SYSTEMS

Low pressure systems mean windy and wet weather. Air moves in towards the centre of the low from areas outside, following the line of the isobars but moving slightly in to the centre. Winds move anti-clockwise around a low pressure system in the Northern Hemisphere. In a deep depression isobars are closely packed together, indicating strong winds. The map indicates the winds will be light becase the depression is not particularly deep.

THE WARM FRONT

A warm front is where warm air meets cold air lying ahead of it as the system moves towards Britain. As the warm front passes over land there will be a period of rain which may last several hours or even all day. This is because the warm air behind the front is forced to rise over the heavier cold air. Behind the warm front is the warm sector. Here the warmer air is being forced to rise up over the cold air ahead of it and clouds form – it may deliver rain or may be more clear and showery. Note on the map that the change of wind direction at the front is indicated by the change in direction of the isobars.

THE COLD FRONT

This marks the leading edge of faster-moving cold air, which is driving into the warm sector and forcing it to rise. Here clouds form to a great height and deliver a short but heavy bout of rain. Behind the cold front, clearer colder air is found, giving sunshine and showers or clear skies. In winter, these showers may be snowy because at each front the winds switch direction and the map shows the winds are now coming from the north so they will bring cold conditions.

INVESTIGATING WEATHER AND CLIMATE *

WEATHER PLOTS

On complex weather maps and synoptic charts, codes are used to plot the weather at a weather station (for example at an airport or weather centre). These provide a summary of the weather at a glance provided that you can work out the key that cracks the code.

GET ACTIVE

Study the weather plot symbols that are used on weather maps at the bottom of the page. Examine each one and think of the weather conditions it points to on a map.

ASSESSMENT I

1. Look at the three weather plots below and and summarise the weather being experienced at each, using the weather plot symbols at the bottom of the page to help you. **(3 marks)**

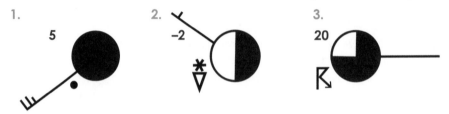

2. Draw a weather plot that shows the following weather conditions:
   - The wind is northerly, blowing at 15 knots. The temperature is –6°C and there are snow showers. There are 7 oktas of cloud.
   - The sky is obscured. There is fog. There is a light wind blowing from the south-east at 2 knots. The temperature is 4°C.
   - The wind is south-westerly blowing at 40 knots. The sky is completely covered by cloud and there is rain and drizzle. **(12 marks)**
3. Study the weather map showing the weather on 1st February 2011.
   - Summarise the weather at each of the weather stations **a**, **b**, **e** and **g**.
   - Describe the weather at station **h** and explain why the weather is like that.

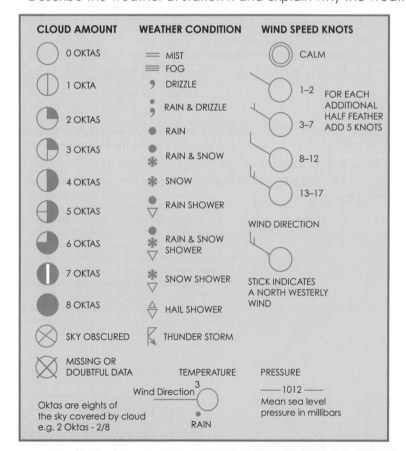

70

INVESTIGATING WEATHER AND CLIMATE *

CO-OPERATIVE LEARNING: TEAM GAME TOURNAMENT

Social skills: coaching others to help learning, competing with other teams.

Learn the weather symbols and as much as you can about weather charts as the topic for the tournament. Get your teacher to devise a series of quiz questions set at different levels, for example some with more help than others or where some people are allowed to use the weather symbol key while others have to remember the symbols.

STARTING OFF

In a team game tournament you should form a **group of four**. You should first learn a piece of work together as a team with everybody helping each other. You should read together, discuss the topic, learn sections of the work in chunks together, give each other little tests by firing out questions so that you are confident you know your stuff. The success of the team depends on everone in the group doing their best to learn about the topic.

LET THE TOURNAMENT BEGIN!

Now the group should **split up** and **each member** should move to another table to join a test group. On the table at this group, each team member will find a test. Some tests may be difficult while others are more straightforward. Your strongest performers should take the toughest test. Each team member now completes the appropriate test. The tests are scored by marking each others' papers at the test table, using given answers.

Each person is given a **score** in the tournament. Now return to your original team **carrying your score** with you. Add all your **group scores** together and enter the total onto a tournament score board. The team that gets the highest score is the tournament winner.

A series of tournaments would let you form a **league** to track team performances over time. You could arrange a transfer system that moves strong players to lower performing teams to boost their league position and to give everybody a chance of being at the top of the league.

SPECIAL SECTION: WEATHER INVESTIGATIONS

HOW TO CONDUCT A WEATHER INVESTIGATION

 **CO-OPERATIVE LEARNING**

1. Individually revisit the 'Investigation Planning Matrix' in chapter 2 (page 37).
2. 'Pair-share' and then 'Round the table' with each person speaking once then anybody can speak.
3. Discuss this issue: What are the main things that go to make good investigations?
4. Lay out the main points your team agree on a piece of A3 paper and post it on a nearby classroom wall.

 **INVESTIGATION ACTIVITY: ASSESSMENT K**

A popular activity in Geography is to conduct a weather investigation. This will involve collecting and recording weather data, presenting the collected data in the form of graphs, tables and charts, and writing about the weather over a period. It may also involve writing about the way the weather has affected people.

COLLECTING AND RECORDING WEATHER DATA

Day/date	Maximum temp (°C)	Minimum temp (°C)	Average temp (°C)	Precipitation (mm)	Weather type	Cloud cover Oktas	Cloud type	Pressure (millibars)	Wind speed	Wind direction	Weather description
1											
2											
3											
4											
5											
6											
7											
8											
9											
10											

INVESTIGATING WEATHER AND CLIMATE *

The spreadsheet at the bottom of page 72 allows you to collect weather data and to record it over a period of 10 days. This can be used later to summarise the weather experienced over the recording period. You should also **collect any news stories** that relate to how the weather has affected people during the period that you are recording it.

Create a spreadsheet like the one on page 72 that includes all the weather elements you intend to collect data about. (The more information you collect about the weather, the more detailed your weather report will be).

SOURCES OF INFORMATION
You can collect weather facts and figures from a wide range of sources, for example:
• weather sites on the internet (Met Office, BBC, etc.)
• newspapers
• TV forecasts
• radio forecasts (Radio 4 Breakfast Programme, Shipping Forecast, etc.)
• satellite images (internet, Met Office, Dundee University Physics)

Be sure to collect information each day to get a complete set of figures. Try to get the information at the same time of day from the same sources. You could design and make your own weather recording instruments or use thermometers and household barometers to take readings and record them. You may have access to school weather instruments or a computerised weather station in your Geography department. One other thing you should do over the recording period is to collect weather stories that show how the weather has affected people as you have been recording it.

PROCESSING AND PRESENTING WEATHER DATA IN A REPORT
This section is designed to help you put together a step-by-step weather investigation. You can follow the steps to produce a standard weather investigation, but if you do some more work on it you can produce something quite spectacular from this template. What you need to do to gain top marks is to write at length using the evidence you have collected to back up what you say. You should try to find local weather averages from a local weather station to see how your data compares to the expected weather for the month you are operating in.

To summarise: take your data, produce an apropriate graph or graphic to show it off and then write about it in geographical terms.

STEP 1: COVER DESIGN
Design a front cover on a sheet of A4 plain paper (if that is the format you are working in). Your front cover design should be neat and tidy and should aim to catch the reader's eye. Use graphics or photos relating to the weather and invent a snappy title in an eye-catching font. You could follow this with a contents page.

STEP 2: AIM AND INTRODUCTION
Write out the aim of your study. Try to tell the reader what you intend to do and why you are doing it. You could introduce the topic by discussing the importance of weather study and how it can help us in our daily lives. Finish this section by telling the reader your method or how you went about collecting the data you will be using in the report.

At this point, you could make reference to the average weather expected in your area.

WRITING THE BODY OF THE REPORT
The next thing to do is to take each section of data, turn it into a graph or graphic and write about it in an analytical way. Give each section of data a heading that will be a topic or chapter heading in your work

INVESTIGATING WEATHER AND CLIMATE *

Temperature

Using graph paper, or a grid on your report paper, draw a graph that shows the maximum, minimum and average temperatures. Temperature should be shown on a line graph because it is time-series data and one day's temperature leads into the next. Annotate your graphs and label the scales appropriately with days and figures in °C.

Then analyse your graphs in a piece of text, pick out the highs and lows, work out the **range of temperature** over the recording period by subtracting the lowest temperature from the highest and comment on the amount of variation there was. Comment on the highest and lowest single daily range. Compare your findings with the local average for the month and comment on whether temperatures were higher or lower than normal. Try to explain why the temperatures were the way they were. Show relationships between clear skies and colder nights for example or high pressure and high or low temperatures depending on the season. Relate temperature to other weather elements such as rain or wind (wind chill).

Precipitation

Precipitation or rainfall is presented by using a bar graph showing rainfall totals for each day. This mirrors the way in a climate graph that monthly averages are shown. The bar graph is used because precipitation/rainfall is unpredictable over a given period. One incident of rain may bear little or no relation to another.

Analyse rain by looking at total rainfall and at the type of precipitation. You may be discussing rain, hail, sleet, snow, frost and mist. Try to match precipitation to other weather elements such as wind and temperature. Cloud types should match up with wet and dry periods. Separate explanations may be given for incidences of fog and mist. One explanation of fog might involve a weather feature called a **temperature inversion**. Rainfall can have dramatic effects on people such as flooding. Snow can provide dramatic stories as can a sudden thaw.

74

Cloud cover

Cloud cover estimations are given in oktas. One okta is one-eighth of the sky covered by cloud. Presenting cloud cover in oktas is best done by dividing a circle into eigths using special symbols. These are special forms of pie charts.

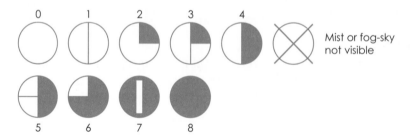

Use a cloud chart from the internet or from a text book to spot different types of cloud. Try to work out what sorts of cloud goes with particular weather. For example, recognise large cumulonimbus clouds that produce thunderstorms. Report on your cloud observations once again, linking them to other weather obsevations. For example, was a cloudy night warmer or cooler than one with less cloud?

INVESTIGATING WEATHER AND CLIMATE *

Atmospheric pressure

Atmospheric pressure varies from hour to hour on any given day. Pressure is relative but as a rule of thumb anything over 1000 mb is high pressure and anything under 1000 mb is low pressure. Note pressure trends from day to day. As the pressure falls, the weather is more likely to become wet, windy and even stormy. As the pressure trend rises, the weather is likely to become sunny and clear. High pressure in winter leads to cold, frosty and wintery conditions. High pressure in summer is often associated with fair weather. Low pressures usually bring mild wet conditions and strong winds in any season. Relate pressure differences in your report to changes in the weather.

MAKE THE LINK

Investigations are a good opportunity to apply what you have already learned in subjects like Maths and English as you put into practice literacy and numeracy skills.

Wind speed and direction

Wind speed is measured in knots (kt). One knot is one nautical mph. We can use the Beaufort scale to give the wind a force, such as gale force 8 or storm force 10 or even hurricane force 12. Copies of the Beaufort scale can be found on the internet to help you with your description of wind and its effects on the environment. Wind direction is an important factor in determining the sort of weather we get. The wind tends to bring the weather conditions with it that were found in its source area, thus, the north wind originating in the polar regions tends to be cold. The south-westerlies in Britain come across the warm Atlantic from Florida, the Caribbean and the Gulf of Mexico, so they bring mild conditions. Okta circles with wind direction sticks attached show the direction the wind came from.

Write about the variation in wind speeds over the recording period. Note how wind and rainfall relate to each other. Note how wind direction can affect the type of weather experienced.

Weather and people

At this stage go over any news stories you have collected and try to show how they related to your own observations of the weather.

CONCLUSION

The final section of the report is a conclusion. In this you can produce a brief summary of your main findings. You might make general comparisons to the average weather for the time of year. Highlight any unsusual weather and summarise the efects the weather had on people. Indicate if some things that happened might be able to be prevented in the future, for example maybe the weather caused flooding that might be prevented by a flood protection scheme. Write out any ways you would improve your work if you were doing the exercise again. Point out any particular successes you had and any things that did not go so well. Round off your report with a good closing statement.

LIST OF SOURCES

Be sure to list all the sources you have used in your research. Include all books, media sources, and websites that you have found useful.

CLIMATE CASE STUDY: FOLLOWING UP ON A STORY

GET ACTIVE   ASSESSMENT J

EAST AFRICA'S CHILDREN'S CRISIS
August 2011

The children of Kenya, Somalia and Ethiopia urgently need our help. Due to a deadly combination of drought, rising food prices and conflict, over two million children under five are currently at risk of starvation in the region.

Right now, we are scaling up life-saving efforts by providing foof programmes and access to safe water and sanitation. We are the main provider of ready-made, therapeutic food for children facing starvation in all three countries, but we desperately need your support to help us reach every child.

A gift of £30 could supply 90 sachets of special high-calorie food. Please donate now.
(Source: UNICEF appeal www.unicef.org.uk)

The appeal above relates to the 2011 famine crisis in East Africa. Throughout the summer of 2011 a looming crisis was developing in this region as the extracts below from a range of internet sources show.

SOURCE 1

Tens of thousands of people are fleeing drought and famine in Somalia in search of food and water in refugee camps in Kenya and Ethiopia. The crisis has been brought on by a deadly combination of severe drought, with no rain in the region for two years, a huge spike in food prices and a brutal civil war in Somalia, where it is too dangerous for aid workers to operate. Somalians are walking as far as 50 miles to reach the Dadaab complex in eastern Kenya, the largest refugee camp in the world. The trek can take weeks through punishing terrain, which is desolate except for the animal carcasses that litter the land.

SOURCE 2

The Australian Red Cross said the organisation had been delivering water and sanitation programmes to drought-ravaged communities in Kenya and Somalia in the wake of the worst drought in 60 years.

Earlier the United Nations (UN) said that more than 11 million people in Africa needed urgent assistance to stay alive as they confront the growing problem of drought.

INVESTIGATING WEATHER AND CLIMATE *

 **INVESTIGATION TASK**

Imagine you are a newspaper columnist and your editor has given you an assignment to follow up this crisis and investigate how it developed. Research the following topics and write a newspaper article with photos and/or graphics about the East African drought and its effects on people. Take notes from the internet sources you use, but remember: do not cut and paste internet source material into your article.

RESEARCH QUESTIONS

- Why does this part of Africa suffer from drought?
- Is global climatic change (global warming) involved?
- How do people in the area make their living?
- What is the scale of the problem in terms of numbers and statistics?
- How does the drought affect them?
- What do drought-affected people do?
- What problems confront the people involved in the drought?
- How are people helped?
- How effective is the help made available?
- What can we do to help?

PEER ASSESSMENT

You are the **newspaper editor**, take an article from somebody else in the class and read it. Check it for mistakes in spelling and grammar as well as for factual errorrs. Meet with the journalist who wrote the article and discuss it, recommending any changes and corrections.

REVISIONS

Redraft the article and print it then post it on the classroom wall.

GALLERY TOUR

Everybody in the class should now tour the gallery of articles on the classroom wall and choose their favourite article. Vote for the article you most admire by placing a sticky dot on it. Take the three most voted-for articles and as a class decide which one should get the Pulitzer Prize for journalism. You could appoint an independent panel from within the school to do this once you have chosen your candidates for the prize.

CLIMATES OF THE WORLD AND THEIR NATURAL ENVIRONMENTS

Learning intentions: in this section you will learn about world climates and the ecosystems they support. You will look into the effects that people can have on the natural environment and how this can cause conflict. You will learn about sustainability and appreciate its importance in helping to protect the environment.

The world has a wide range of climates. Some areas are hot and wet with over 2000 mm of rainfall, for example the equatorial regions with their rainforests. Other parts of the world are hot but dry, getting less than 250 mm of rainfall each year (the hot deserts). As a complete contrast the cold regions, such as the Polar lands where temperature is low all year, reaching extremes in winter, have low precipitation (even below 250 mm), making them cold deserts. Huge areas of the Northern Hemisphere are covered in coniferous forests: these are called the Taigas. They experience continental climates, with extreme cold in winter but warmer conditions in summer, allowing trees and other vegetation to grow over the short growing season. One of the key climatic regions of the world are the great temperate grasslands of Asia and North America, which lie between the Tropics and the Taigas and produce much of the world's grain harvest. Another of the hot climates is the savanna climate typical of large parts of Africa: this is tropical grassland.

78

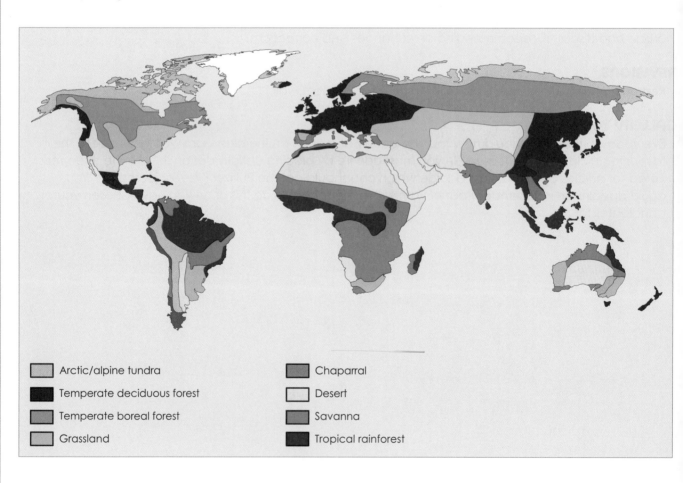

▢ Arctic/alpine tundra		▢ Chaparral	
▢ Temperate deciduous forest		▢ Desert	
▢ Temperate boreal forest		▢ Savanna	
▢ Grassland		▢ Tropical rainforest	

Britain's climate is a cool north-west European maritime climate. We often take our summer holidays in areas that have a Mediterranean climate with hot dry summers and mild moist winters.

BIOMES

A biome is a large geographical area in which animals and plants are adapted to the local environment. The climate and physical geography of these areas determine what type of biome is found. Major biomes include deserts, rainforests, temperate grasslands, savanna grasslands, tundra and taiga. Biomes have specially adapted ecosystems involving all plant and animal life.

1. Use an atlas to investigate world climates and biomes. The beginning of an atlas often has a reference section that will help you do this.
2. Compare world climates shown on one map with world vegetation zones on another. What do you notice about these two maps?
3. Study detailed maps showing winter and summer temperatures, patterns of rainfall and atmospheric pressure patterns.
4. **Task:** make some short bullet point notes that summarise your observations about climate, weather and vegetation patterns on the atlas maps you have studied. Can you write out some explanations of the patterns shown on these maps?

MAKE THE LINK

In Biology we study ecosystems and the plants and animals that live in particular parts of the world or the seas around us. Geographers and Biologists play a key role in conservation. They often work for conservation bodies, for national park authorities and as ecologists in the business world. All large-scale civil engineering projects, such as the Edinburgh Tram, the Edinburgh Airport development plan, or major road schemes around Glasgow, employ environmental consultants. Sustainability is the key concept in environmental protection.

PEOPLE AND NATURAL ENVIRONMENTS

Learning intentions: in this section you will develop your presentation skills, which will enable you to put on a presentation for an audience. This will help you to learn about a natural environment and its environmental issues.

Social skills: preparing with the audience in mind, speaking with confidence and sharing ideas with others.

CREATING A PRESENTATION USING POWERPOINT

If you are preparing a presentation for your class, software like Microsoft™ PowerPoint is an effective tool to use. You need to be careful though: lots of people use this piece of software and don't always do it well. The first thing to remember is that PowerPoint is designed to illustrate a presentation. The presentation itself is all about interacting through speaking with an audience. Presenting the findings of an investigation to others is a key geographical skill.

RESEARCHING PEOPLE AND NATURAL ENVIRONMENTS

The best way to go about this research-based project is to think of it as a TV programme like *Living Planet* or *Coast*. What the presenters of these programmes do is interest the audience and create a joined-up narrative that takes the viewer on a journey of discovery with them.

Your programme will need an intriguing topic, an interesting introduction that gives a flavour of what the programme is about to reveal, a body of information that is well linked together and finally a conclusion that gets the viewer thinking about what has gone before and reinforces the argument of the programme. Your presentation should aim to have such a structure, and in place of moving TV images you should use PowerPoint slides to illustrate your main points and provide interesting visual references.

The Arctic and Nunavut

GET ACTIVE

INVESTIGATING A BIOME AND THE IMPACT OF CHANGE: ARCTIC LANDS

Your task is to investigate the people and wildlife that live in the spectacular environment of the tundra biome. You will need to use the internet to research some of the questions posed here. You should also try to make judgements based on evidence presented to you, such as in data tables and photographs.

INVESTIGATING THE CANADIAN ARCTIC NUNAVUT

Use the internet and any other sources to find out about this region. Make a start by accessing Wikipedia.

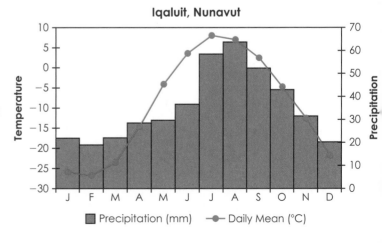

Iqaluit, Nunavut

Precipitation (mm) — Daily Mean (°C)

80

LOCATION AND CLIMATE

1. Where is Nunavut? Use the map on page 80 to help you describe where Nunavut is. Describe its position in relation to the Arctic Circle, Canada, Alaska, and Greenland.
2. Why and when was it created?
3. Use the climate graph for Iqaluit to describe the climate of this area. Refer to the temperatures shown, calculate the range of temperature, and note the highest and lowest temperatures. Show how many months have a temperature below freezing. Calculate the total rainfall (precipitation). Work out how many months during which it is likely to snow.

Using photos and media as sources

1. Describe the scene in photograph 1, which shows part of the Arctic. In your description, use terms like 'in the foreground' and 'in the background'. Be sure to include everything you can see.
2. With a partner, compare your written description with theirs. Give constructive criticism of their work.
3. Draw a sketch of the first photograph and label it to say what you can see. Annotate the photograph to show reasons why this would be a difficult place to live.
4. Study photographs 2 and 3. Explain, using a spider-diagram, why it is difficult to travel in the Arctic.
5. Watch a video or film clip about the Arctic and make notes for your investigation.

VEGETATION AND WILDLIFE

1. What is the natural vegetation like in Nunavut and how is it adapted to the tundra climate?
2. Find out about a selection of wildlife found inthis biome. How do they adapt to this natural environment?

- Arctic Fox
- Snowy Owl
- Seal
- Ptarmigan
- Polar Bear
- Caribou
- Musk Ox

PEOPLE

1. Who lives in Nunavut? Find out about the inuit and their way of life. Find out about other people who have come to make a living here.
2. How do people manage to adapt to life in this area?
3. How has life changed in modern times compared to the way it was in the past?

ENVIRONMENTAL CHANGE

1. Research how this environment is undergoing change. Draw up a table like the one below and summarise your research.

Environmental change or development	Impact on the tundra

2. Write about the way new developments (such as the search for oils and minerals) are likely to have an impact on the tundra biome.
3. Do you believe that this delicate environment should be specifically protected? Give reasons for your answer.

GET ACTIVE — OTHER STUDIES OF NATURAL REGIONS

Investigate other natural regions. You can design similar questions to those above, just substitute your own natural region, wildlife and people for the tundra examples used.

EQUATORIAL RAINFORESTS: THE RAINFOREST NATURAL REGION

Equatorial rainforests circle the globe, following the line of the Equator (zero degrees of latitude). Here you will find the **equatorial climate**, which extends roughly 6° north and south of the Equator. Other rainforests can be found in the tropical zone lying between the Tropic of Cancer and Tropic of Capricorn, provided there is sufficient rainfall for a large part of the year.

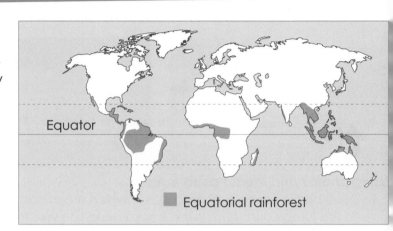

Equator

■ Equatorial rainforest

1. Find a blank world map image on the internet.
2. Insert it as an image into a drawing program such as 'Paint'.
3. Use the line tool to mark on the Equator and the two Tropics.
4. Use the drawing tools to paint in the world's equatorial rainforests and label the diagram with parts of the world and the countries where rainforests are found.

EQUATORIAL CLIMATE

The climate at the Equator is always hot, on average the daily temperature is in the region of 26°C. It is even warm at night. The high average temperatures are matched by constant high humidity and heavy rainfall. This makes the equatorial region a difficult climate to live in. Rainfall totals exceed 2000 mm per year and can be over 10 000 mm per year in some areas, for example some parts of the Amazon Basin.

Study the climate data in the table at the top of page 83 for Manaus in the Amazon rainforest of Brazil.
1. Calculate the average daily temperature and enter it into the table.
2. Calculate the total rainfall for the year and enter it into the table.
3. Calculate the total number of days and work out what percentage of the 365 days in a year are rainy humid days.
4. Using Microsoft Excel™, enter the statistics above into a spreadsheet and create a bar graph to show rainfall, and line graphs to represent temperature. Choose an appropriate form of graph to represent rainy days.
5. Describe the climate of Manaus Brazil in a single paragraph by summarising the weather pattern shown above.
6. Write a summary of what **a day's** weather in the Amazon might be like for someone like Mungo Smith working in this environment (see the extract from Smith's Journal on the following page).
7. Why would it be difficult to live and work here in the Amazon?

CLIMATE OF MANAUS, BRAZIL

Month	Minimum average temp (°C)	Maximum average temp (°C)	Average daily temp (°C)	Rainfall (mm)	Rainy humid days
Jan	24	31		250	20
Feb	24	31		230	20
Mar	24	31		260	20
Apr	24	31		230	18
May	24	31		170	18
Jun	24	32		90	16
Jul	24	32		60	8
Aug	24	33		60	6
Sep	24	33		60	8
Oct	24	33		130	10
Nov	24	33		141	12
Dec	24	32		230	16
			Year average	Total/Year	% rainy days per year

It was by now mid afternoon and the giant clouds were building as I surveyed the great sea of forest crowns from my Amazon treetop vantage point. The burning sun of the mid morning was at last disappearing behind the bubbling cumulo-nimbus thunderheads, offering at least a little respite from the uncomfortable heat and humidity. Starting with a few large drops pitter-pattering on the tin roof of the treetop observation platform, the rain burst into torrents and a spectacular show of lightning and roaring thunderclaps brought the usual afternoon soaking. The winds now were getting up and the treetops moving in their own rhythm left me uneasy. It was time to slide down the ropes and return to the village, wildlife study was over for the day. The sounds of monkeys, birds and jaguar had now stopped and only running water could be heard drowning out everything else as the rain battered off the canopy and its leaves. The rain would be on for an hour or two at least but it was pretty certain that the sun would be out again later with the prospect of another warm humid evening in store. (Journal of Mungo Smith, Amazon Ecologist)

RAINFOREST ECOSYSTEM

The rainforest ecosystem is one of the most diverse in the world. This means that the rainforests are home to more species of animal, bird, insect, fish and other creatures than any other of the world's natural regions. Here are found a vast number of the world's plant species and a massive percentage of the world's trees. Many people believe that the rainforest plays a key role in keeping the climate of the planet stable and has an effect on the oxygen/carbon dioxide balance in the atmosphere. In the last hundred years large areas of the world's rainforests have been disappearing due to forest clearance and fires started by people or due to drought brought on by climate change.

RAINFOREST DESTRUCTION: AN ENVIRONMENTAL ISSUE

WHAT IS AN ENVIRONMENTAL CAMPAIGN?

Below is an appeal released on the internet to try to stop the Belo Monte Dam being built in the Brazilian Amazon. This is a case of environmental conflict, where various groups of people have very different points of view about an environmental issue. By the time you read this it may be too late for you to take part in the petition if you feel strongly about the case being put, but there is no reason for you not to investigate this conflict and try to understand the different points of view involved.

Rainforest destruction

> On behalf of the Juruna Indigenous people of the Xingu River Basin, I ask for your support to help stop the Belo Monte Dam. At any moment, the Brazilian government could break ground, causing irreparable impacts for our communities, the environment and the global climate. We are at a critical time in the campaign to stop the Belo Monte Dam and it is essential that the international community take action now to defend the Amazon and support indigenous peoples' rights.
> **Please help us defend our river and future generations by signing this petition.**
> *Sheyla Juruna, Juruna Tribal Leader*
> (Source: http://amazonwatch.org/work/belo-monte-dam)

THE BELO MONTE DAM PROJECT

The Brazilian Government is continuing to develop its Amazonian Region and sees this as an important part of the country's overall economic development strategy. Brazil is a growing industrial power and is hungry for resources to supply its industry. The resource-rich Amazon Basin is seen as a key part of the country's future plans. Brazil also needs to produce power resources and has built a number of large dam projects in the Amazon Region to generate hydroelectric power. The Xingu River in the state of Pará is a tributary of the Amazon and the government of Brazil wishes to construct the world's third-largest dam across this river. This has lead to two decades of national and international protest in an attempt to prevent the project being started because of the environmental damage it will cause.

RESEARCH PATH: ASSESSMENT H (20 MARKS)

Follow this research path to gather as much background information as you can to investigate the issue of the Belo Monte Dam.
1. Route 1: Belo Monte search on Wikipedia.
2. Route 2: Research Belo Monte more generally on the web, for example at:
   • www.internationalrivers.org
   • www.guardian.co.uk/environment
   • www.survivalinternational.org
   • www.bbc.co.uk/news

3. Route 3: Write to the Brazilian Embassy in London requesting information about the Amazon and the Belo Monte project.
4. Route 4: Investigate one of the Dam's biggest investors, Brazilian mining giant Vale (www.vale.com)

What conclusions have you come to?
• Summarise the points of view each of your sources take.
• Comment on how good the point of view is. Is it reliable? Say why you think so.
• What do you think should happen with regards to this project?
• Why should rainforests be specially protected from developments?
• How can rainforest resources be exploited in a sustainable way for the benefit of local people?
• What can you do about this rainforest issue?

ASSESSMENT K

Look at the diagrams.

1. Complete the table below by adding the appropriate name, letter or graph number. **(2 marks)**

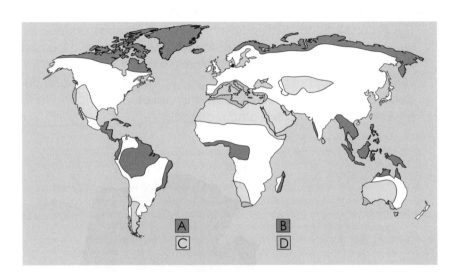

Letter and name on map	Graph number
A	
B	
C	
D	

Choose from **tundra, hot desert, rainforest, mediterranean.**

2. Describe, in detail, the main features of the climate shown in Graph 4. **(3 Marks)**

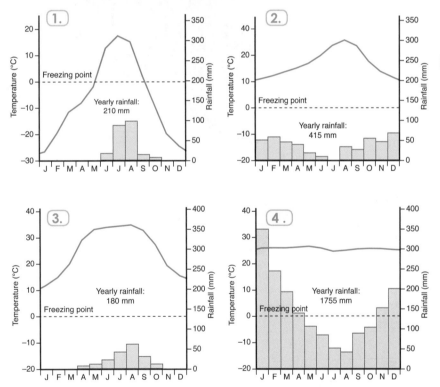

85

Form a team of three or four. Discuss the questions above and create a shared answer plan that gives clear instructions to a marker as to how the marks will be awarded for the questions. Use the answer plan to mark somebody else's question. Look at your own marked answer and analyse how well you did. What could you do to improve your answer? Discuss this in your group.

HOW DO WE HOLD AN ENVIRONMENTAL DEBATE?

ARCTIC NATIONAL WILDLIFE RESERVE

The Arctic National Wildlife Reserve is a hotly contested issue in the United States of America. The reserve is the world's largest Arctic conservation area, it lies on the North Slope of Alaska and along the shores of the Arctic Ocean. It was first designated for special conservation back in the 1950s and 1960s but since then huge oil reserves have been discovered in the area and national debate between conservationists and developers relating to the impact of oil development in an environmentally sensitive area is ongoing.

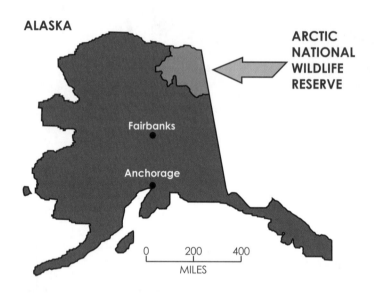

ORGANISING A CLASS DEBATE

This is an ideal issue to debate. It requires students to get involved in role play where they represent some of the opposing parties interested in conserving or exploiting the Arctic National Nature Reserve.

ROLES

The class should decide who will chair the debate and who will support the chair. You will need a team of Chair Person, a Recorder and a Reporter. You will also need **six** teams of **four** students to research and debate the issue

ISSUE: SHOULD OIL DEVELOPMENTS BE ALLOWED IN THE ARCTIC NATIONAL WILDLIFE RESERVE?

The roles of people in the debate are laid out below. One/two periods should be devoted to researching the arguments in the debate. One period should be given to debating the issue. One period should be given to presenting the findings of the debate in a report by the Chairman's team.

MAKE THE LINK

Debating is a key part of politics. Both the UK and Scottish Parliaments debate issues. In Modern Studies and English, debating is an important skill. Parliamentarians and politicians around the world work with similar rules of debate that you will use here.

Rules of debate

1. The Chair Person or one of the team keeps order, controls the debate, decides on points of order and controls points of information.
2. All speakers speak through the Chair and recognise its authority.
3. Teams are allowed two speakers, the main speaker and the support speaker.
4. Researchers can assist the speakers but cannot speak until the debate is opened to the floor.
5. Speakers are called in turn by the Chair in the order: main speaker **for**, main speaker **against**, support speaker for, support speaker against, until all speakers have been heard.
6. Main speakers speak for **four** minutes, support speakers speak for two minutes. The Reporter indicates when there is **30 seconds** to go.
7. The Chair ends speeches within **10 seconds** of the allocated time by calling time on the speaker. At this point the speaker **must** finish.
8. After all the set-piece speakers have been heard, the debate is opened up to the floor. Each person other than the main speakers may speak once at the chairman's discretion.
9. The last speaker in each team sums up the arguments for their side in **two** minutes.
10. The Chair organises a vote with the Recorder acting as vote teller for one side and the Reporter acting as teller for the other. Voting is by show of hands and a shout of Aye or Nay in favour or against.
11. All those voting should do so on the basis of the strength of the case put in the debate and not on the basis of their team's point of view.
12. The Chair has the casting vote in the case of a tie.
13. The Chair Person's team summarises the debate and presents the findings in a written report to the class.

CONTRASTING COUNTRIES

Learning intentions: in this section you will develop a range of skills that will allow you to compare and contrast countries around the world. You will be able to understand the differences that exist between rich and poor countries and you will be able to find out about the reasons behind these differences. Your starting point will be your own experience of Scotland and you will use Scotland as a benchmark for comparing.

MEASURING THE DIFFERENCE BETWEEN COUNTRIES

The spreadsheet demonstrates, even across a limited number of the countries that great differences are found between them. Governments around the world collect facts and figures about their country that are very useful to the geographer comparing one country with another. We can use various statistics that reveal how well or how badly a country is doing against similar or contrasting societies. We can also try to find out why a particular country seems to be doing better or worse than another in terms of how it is developing. These measuring statistics are called **indicators**.

Country	Crude Birth Rate (CBR)	Cruse Mortality Rate (CMR)	Infant Mortality Rate (IMR)	Adult Literacy	Gross Domestic Product (GDP)	Population total (POP)	Life Expectancy (at birth)
United Kingdom	12	9	4	99	35000	62 M	80
Bangladesh	23	6	53	48	1770	157 M	69
South Africa	20	6	53	86	11000	49 M	49
Ethiopia	43	12	79	43	1000	88 M	55
Brazil	18	6	22	89	11000	202 M	72
Saudi Arabia	19	3	16	79	24000	26 M	74
Egypt	25	5	26	71	6000	81 M	72
Canada	10	8	5	99	40000	34 M	81
Japan	7	9	3	99	34000	127 M	82
Nigeria	36	16	93	68	2000	52 M	47
Pakistan	25	7	65	49	2000	185 M	65
France	12	9	3	99	33000	64 M	81
South Korea	9	5	18	86	30000	49 M	79
Mexico	19	5	18	86	14000	113 M	76

INDICATORS OF DEVELOPMENT

The spreadsheet shows the variations in development and health between developing countries, and the contrasts between the poor less-developed world with the rich developed world.

Before going on to investigate these differences it is important to define our terms. Knowledge of the correct words and what they mean helps with more detailed and accurate analysis of development issues.

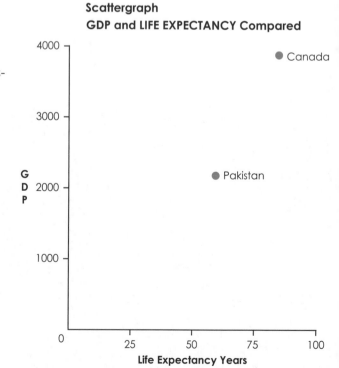

Scattergraph
GDP and LIFE EXPECTANCY Compared

A WORLD OF DIFFERENCE: CONTRASTING SOCIETIES

COMMON DEVELOPMENT GEOGRAPHY TERMS

- **Development** – this refers to social and economic development and can be measured using indicators of development. Linked terms may be 'more economically developed countries' (MEDCs), and 'less economically developed countries' (LEDCs).
- **Capitalist market economies** – the group of the most economically developed countries such as the USA, the EU states, Japan, Canada, Australia, etc. These are free market economies and are under democratic control.
- **First World** – the most developed countries. See Capitalist market economies above.
- **Third World** – the developing countries.
- **Newly industrialised countries** – refers to countries in the developing world that have seen industrialisation expanding as part of their economy rather than their economy being dominated by primary industry such as agriculture. Examples might be found in parts of south-east Asia such as Malaysia or South Korea.
- **Least developed countries** – the group of the poorest developing countries. The sub-Saharan countries of Africa such as Niger and Chad are examples from this group alongside poor Asian countries such as Bangladesh and Nepal.

INDICATORS

A number of measures of relative development can be used to compare and contrast different societies. These are called indicators of development; they may be economic indicators such as **gross national product (GNP)**, health indicators such as **infant mortality rates (IMR)** or social indicators such as levels of **adult literacy**. Sometimes these indicators are grouped together in groups, such as the **Physical Quality of Life Index (PQLI)** or the **Human Development Index (HDI)**.

HUMAN DEVELOPMENT INDEX (HDI) – 2010 RANKINGS

Highest rankings	Country	Lowest rankings	Country
1	Norway	129	Mali
2	Australia	130	Burkina Faso
3	New Zealand	131	Liberia
4	United States	132	Chad
5	Ireland	133	Guinea-Bissau
6	Liechtenstein	134	Mozambique
7	Netherlands	135	Burundi
8	Canada	136	Niger
9	Sweden	137	Congo
10	Germany	138	Zimbabwe

GET ACTIVE

Why not get a fundraising event going in your school? you could target your efforts at providing antimalaria bed nets, supporting a school in another country, providing disaster relief, fighting AIDS or supporting a charity such as the Red Cross.

OUR EVERYDAY LIVES

The news is often concerned with events in poor countries. Major events are organised using TV campaigns such as Red Nose Day (Comic Relief) or Children in Need that raise money and help people in many countries around the world.

GETTING TO GRIPS WITH INDICATORS

WHAT DO ECONOMIC INDICATORS MEAN?

Economic indicators such as **Gross National Product (GNP)** have been used widely in the past to compare countries. GNP is defined as 'the total value of all goods and services produced by a country'. It includes investments and government expenditures. It is calculated per year and may be divided by the total population to produce the figure of GNP per capita (per head).

Gross Domestic Product (GDP) per capita is another common economic measure. It is calculated by dividing the money value of all the goods and services produced by a country in a year by the total population. This value ignores the so-called invisible earnings in GNP (mostly investments). You may also come across **national income per head**, which is a measure of how much people earn per year.

CAN WE RELY ON ECONOMIC INDICATORS?

As with all indicators, you should use caution in relying on them. They don't always give the whole story.
- GNP and GDP are most accurate as measures of development where there is a lot of business going on – lots of transactions.
- In some parts of the world goods are bought and sold at local level using a bartering system. This business is, by its nature, difficult to estimate or record so it may not be included in the figures.
- Some economies are dominated by subsistence farming. Here, both farmers working locally and self-sufficient communities do not contribute to the 'measurable economy' but may well be thriving.
- National figures for GNP or GDP may mask considerable differences that exist within a country. For example, there are great differences within the economy of Brazil if you compare its highly populated urban scene with the relatively undeveloped Amazonia Region.
- There may be great divisions within societies in terms of wealth. For example townspeople may be wealthier than countryside dwellers. Take the example of the economic differences between black and white societies in **apartheid** South Africa. These still persist today.
- Exchange rates vary from day to day and year to year. This can distort the real value of trade so special adjustments have to be made to take this into account if valid comparisons are to be made. Check € (Euro) against £ (sterling) in a newspaper or on the internet and see how the rate changes over a week.

Economic indicators do form a sound and reasonably accurate measure of the development of a nation. Their deficiencies are lessened when other social and health indicators are used alongside them.

Use the spreadsheet on page 88 to complete the following tasks:
1. Draw a bar graph to show differences in GDP between the countries in the spreadsheet.
2. Describe what the bar graph shows in a paragraph of writing. Refer to high points and low points. Calculate an average GDP for the group and compare individual countries to the average figure.
3. Draw a line graph with two axes. Make the vertical axis **GDP** and the horizontal axis **life expectancy**. Plot the figures from the spreadsheet that show GDP and life expectancy on the graph. Draw a best fit line that creates a graph linking all the dots in a line or curve.
4. What is the relationship between GDP and life expectancy? Explain it in your own words.
5. Why are GDP and life expectancy linked? Give reasons to support your answer.

A WORLD OF DIFFERENCE: CONTRASTING SOCIETIES

WEB SEARCHES

Web searches can provide lots of statistical data relating to world economies and economic indicators. You should try searching for these sources using a standard search engine such as, Yahoo, Google or Bing. Sites such as Eurostat, CIA World Factbook and the Population Reference Bureau can also be useful in your research.

SOCIAL INDICATORS

Typical social indicators used in development studies are: crude birth rate, crude death rate, infant mortality rate, maternal mortality rate, access to safe drinking water, adult literacy rate, life expectancy at birth, doctors per patient, hospital beds per patient.

HOW ARE SOCIAL INDICATORS CALCULATED AND WHAT DO THEY MEAN?

Crude birth rate (CBR) is measured by recording all births in a country and it means the number of live births per 1000 head of population in any given year. **Crude mortality rate** (CMR) is measured by recording all the deaths in a country, it means the number of deaths per 1000 head of population in any given year. **Infant mortality** is the number of babies who die within the first year of their lives per thousand live births in a given year. **Adult literacy rate** is a measure that indicates the number of adults, as a percentage of the population, who have a basic primary school education and can therefore read and write at a basic level. It can be a different figure for males and females.

Indicators based on social welfare and health characteristics of countries can be used to measure relative development. Below are a number of obvious statements which you might hear when discussing development issues.

- Developing countries have higher birth and death rates than developed countries. (See the demographic transition model.)
- Infant mortality rates are higher in developing countries than in the developed world.
- There are fewer doctors per patient in the developing world.
- Medical provision is generally poor in poor countries, e.g. fewer hospital beds per patient.

MAKE THE LINK

In Maths and Science we learn to draw and interpret graphs. Interpreting graphs gets us to look closely at figures to see what they show so we can draw conclusions. Graphs are used as supporting evidence when we write and draw conclusions. They are an effective way of using figures and statistics found on the internet.

- People have less access to safe drinking water in **less economically developed countries** (LEDCs).
- Lower levels of literacy are the norm in developing countries.
- Energy consumption is lower in poor countries than in the developed rich world.

Be cautious when you use sweeping statements, they are not always totally correct even if they appear to have some basis in truth. Remember to try to base things you say on evidence provided by careful use of facts and figures.

BE CAREFUL HOW YOU SAY IT

Common exaggerations:
- *People in the third world all have large families.* This is not the same as saying that birth rates tend to be high.
- *The reason that the birth rate in Country X is high is that there is no birth control.* Limited access to birth control may be a feature of many developing countries but that does not mean a total lack of birth control.
- *Infant mortality is high because there are no doctors.* Fewer doctors per patient than in Western Europe does not mean that there are no doctors at all to treat patients.

Be cautious in expressing points of view: be careful to say exactly what you mean. Using facts and figures adds weight to your answers, especially if they come from a reliable source such as the United Nations.

OTHER WAYS OF MEASURING DEVELOPMENT

COMPOSITE MEASURES OF DEVELOPMENT

Rather than rely on purely economic or social measures of development, workers in the field and agencies like the UN tend to join economic and social indicators together into an **index**. One such index is the **Physical Quality of Life Index (PQLI)**; this combines infant mortality, literacy and life expectancy into one composite measure on a 0 to 100 scale. Most developed countries score over 90 while the poorest countries, such as Mali, score under 30.

Human Development Index (HDI) is now the measure favoured by the **United Nations** organisation for making comparisons. This moves away from purely economic measures such as GNP or GDP. These measures show only part of the picture. HDI includes a social dimension to assess the quality of life and not just the standard of living.

HDI is measured on a scale between 0 and 1. It is calculated by taking statistics relating to the following characteristics of a population.
• Longevity measured by life expectancy at birth.
• Income measured by GDP in $US per capita, converted into purchasing power to overcome the problem of exchange rates. This reflects what money will buy at local rates.
• Education measured by the adult literacy rate and school enrolment ratio.

The extract below shows an HDI table from the UN website. It shows how the various indicators are converted and composited to produce a listing in the form of a league table of development, with Norway in top position. Niger in West Africa, with an index rating of 0.340, is found towards the foot of the table. The table also shows how countries can move up and down compared to past performance.

Highest Rankings	Lowest Rankings
1. Norway	128. Central African Republic
2. Australia	129. Mali
3. New Zealand	130. Burkina Faso
4. United States	131. Liberia
5. Ireland	132. Chad
6. Liechtenstein	133. Guinea-Bissau
7. Netherlands	134. Mozambique
8. Canada	135. Burundi
9. Sweden	136. Niger
10. Germany	137. Congo

1. Choose 10 contrasting countries.
2. Create a table in an MS Word document.
3. Choose a short list of indicators such as GDP.
4. Enter the figures in the table.
5. Give each country a rating 1–10 (1 is the richest and 10 the poorest).

Country	Indicator 1	Indicator 2	Indicator 3	Indicator 4	Rank
Scotland					

6. Do an internet search for the latest version of the **Human Development Index** to see how the rankings of different countries have changed. Which countries are in the top 10 and which are the bottom ranked countries? (At http://hdr.undp.org/en/data/map/ you will find the United Nations Development Programme UNDP and all its HDI maps and statistics.)

MAKE THE LINK

In Maths we use indices to help us manage complex figures. An index helps us to make comparisons between statistics that are not all about the same things. For example, in the HDI some statistics are given in US$, others are percentages or are expressed (like crude birth rate) as per thousand. Only by making the figures into an index can we join them together to get a complete picture. These figures can be used to support work in RME and in Modern Studies.

SITUATION

The **situation** of a place is where it sits in the landscape and other features around it. Unlike site arrangements, a town's situation is unique to it so we have to describe it in terms of what is all around it.

FUNCTION

The **function** of a town is what it does to make its way in the world. It is one of the primary reasons for existing and is a key reason to explain why it might thrive and grow or, in some cases, why it might lose its importance as social and economic conditions change.

Defensive sites. In Scotland, some of our main towns and cities started on sites chosen for defence. Both Edinburgh and Stirling grew up around castles on hill tops that provided good defence from enemy attack.

Underground resources such as coal and metal ores allow the creation of mining villages and mining centres. Fife had a number of these towns, such as Cowdenbeath or Kelty.

Some towns have the function of being industrial towns, producing things like iron and steel or textiles, such as cotton or wool. In the past, Dundee produced jute, Glasgow produced cotton. Some Borders towns still have woollen mills.

107

Some towns are important as transport centres, for example, where main roads meet. Examples are Stirling and Perth.

Dry sites and **wet sites**. Towns may grow near to water as they can use the water resource available from a river or nearby loch. These are called water-seeking sites.

Other towns may try to avoid flooding by using a site above the floodplain of a river on a nearby slope. This still allows access to the water supply.

Many towns set up on the spring line (a place where underground water can be lifted up from wells or where water comes out of the ground at a spring). In many cases, the spring line marks the level of underground water at the base of a hill.

MY PATCH, MY TOWN, MY STREET

LET'S GET ACTIVE

Where you live – your patch – is very important to you. It is the centre of your mental map and probably the place you know best of all. Investigating your street, your village or part of your town can be done using a RICEPOTS survey. This is a land-use survey of a particular area, which is very easy to conduct. The best way to start is to create a topological map.

TOPOLOGICAL MAPS

The London Underground map is the best known of all the topological maps. It does not show scale and distance it only gives rough direction and it is mostly concerned with railway lines and stations and their inter-connections.

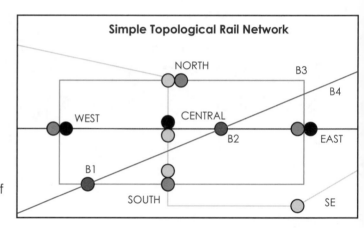

The rail network in the illustration on the right is easy to read and you can use it to plan a journey from one station to another choosing the best route. To create a topological map of your street you can build up a diagram using the drawing tools in a program like MS Word.

In the map each box represents a building on Main Street. None of the buildings are to scale so it is easy to create.

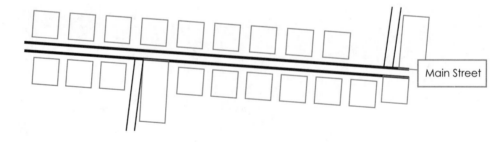

RICEPOTS

RICEPOTS is a way of identifying what each building in an area is used for when a fieldwork survey is done.

Letter	Land Use
R	**R**esidential (houses and flats)
I	**I**ndustrial (factories, workshops and industrial buildings)
C	**C**ommercial (shops, wholesalers, suppliers)
E	**E**ntertainment (cinemas, theatres, sports and leisure)
P	**P**ublic (town hall, schools, churches, museums, libraries)
O	**O**ffice (solicitors, accountants, head office)
T	**T**ransport (rail, bus, airport, taxi)
S	**S**ervice (banks, hospitals, police, fire, ambulance, restaurants, hotels)

The table on the right shows the result of a RICEPOTS survey of Main Street conducted by a school student. She walked along the street and using a pre-prepared map she recorded the land use of each building on the map. The colour code helps identify patterns where certain land uses are grouped together and dominate an area, such as the town centre. Annotations can be used to add other information from the survey.

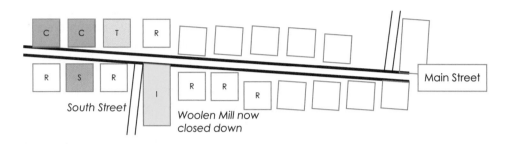

MAP ACTIVITIES

1. Plan a direct route on the rail network map from West Station to SE Station, noting other stations you will pass on the way.
2. Plan an alternative route and do the same for it but try to make it longer.
3. Create a topological map of your route to school, plotting any points of interests such as bus stops and important buildings along the way as little circles with names attached.
4. Draw a topological map of your street or part of your neighbourhood and complete a RICEPOTS survey.
5. Create a colour-coded key for your map.
6. Calculate the percentages of each land use and rank them in order.
7. Draw a bar graph showing the percentages of each land use in the study area.
8. Write a brief report to show your findings and share them with your class.

109

CO-OPERATIVE LEARNING

Social goal: agreeing tasks and dividing up work.

Academic goal: to complete a field study land-use survey and write up the results.

1. Form groups of four made up of students that live near to each other.
2. Create a topological map of your agreed study area on a large piece of paper. Cut this into four sections and take a section each to be completed.
3. Agree to survey a number of local streets with each person doing a RICEPOTS survey on their part of the topological map.
4. Complete the surveys.
5. Rejoin the map or redraw it and colour-code it.
6. Complete steps 5–8 of the **Map activities** task above as a group by sharing tasks.

EDINBURGH

Learning intentions: in this section you will learn about the way cities have problems in enabling people to move about and how transport can have an effect on the quality of life in the city. You should be able to discuss how transport issues may be resolved in the future.

Edinburgh has many challenges because large numbers of people use its streets every day. Tens of thousands of people travel into the city centre from houses on the outskirts of the city. Thousands more travel into Edinburgh to work from Fife across the Forth Road and rail bridges. People travel into Edinburgh from the Borders, from across the central belt of Scotland and from the East Coast towns such as Musselburgh, Dunbar and Berwick-on-Tweed across the English border. People who travel into the city for work are called commuters. They travel by car, bus and train, but also by bike and on foot.

Besides people travelling to work, many come into town to go shopping, to access services such as education, or to use banks and other commercial businesses. People travel in to the city centre for entertainment, visiting theatres and cinemas, restaurants and clubs. Another large group of people visiting the city centre are tourists, who expand the normal city population by hundreds of thousands, especially at Festival time when the streets are packed with tourists and visitors attending events and shows such as the Edinburgh Military Tattoo.

GET ACTIVE

110

1. Use your GIS skills to find a range of maps and satellite images of Edinburgh and Edinburgh city centre. Focus on transport issues, problems and solutions. The table below gives an idea of places to start looking for information.

Map/Image Source	Type of Resource	Potential Use
Google Earth	Map and satellite images street views and photos	Creating maps identifying street patterns and transport pinch points
OS 'Getamap'	OS map and photo resources	Analysing transport patterns
Edinburgh Council website	Aerial photos, information about transport planning and maps	Analysing street patterns, observing traffic densities, provides information about traffic management
Webcams	Live images	Doing remote traffic counts
Travel Information: AA, RAC, Radio Forth local radio, BBC website	Traffic information	Identifying traffic problems when and where they occur
Railway timetables and rail travel information, e.g. Scotrail	Timetables	Indicates rail flows to and from the city and gives information relating to where commuters come from
Lothianbuses.com	Bus timetables and route maps	Reveals flow patterns along bus routes, shows the origins and destinations of comuters
Edinburgh Festival website	General information	
Edinburgh Airport website	Airport arrivals	
Tourist information	General information	

2. Identify in detail some of the traffic management issues affecting Edinburgh by developing a series of spider diagrams that outline the main points. This diagram will give you a start.

BIG QUESTIONS

Use these questions to write about traffic problems in Edinburgh or adapt them to write about problems in another city you have studied.

1. Does Edinburgh have traffic problems?

2. What measures have been taken to deal with them?

3. How successful have the measures been in dealing with the traffic problems?

4. How have other large cities in the UK, Europe and the world dealt with similar problems?

5. What would a plan for Edinburgh's transport future contain within it that will deal with the growing traffic and travel problem?

6. Should Edinburgh introduce congestion charges like London has?

7. What effect will the following traffic schemes have: The Edinburgh Tram, The New Forth Bridge, The Borders Railway Scheme, The Airport Development, The Ocean Terminal development?

ASSESSMENT L

1. For Edinburgh, Glasgow or any city you have studied, choose **one** way that traffic congestion could be reduced:

• Improve public transport ☐
• Put parking measures in place. ☐
• Encourage greater use of cars ☐
• Make people walk to work ☐

(2 marks)

2. How can cycle lanes help avoid traffic jams? **(2 marks)**

3. Why is pollution a problem in the city centre? **(2 marks)**

4. For Edinburgh, Glasgow or any other city you have studied, **describe** possible solutions to the problem of city centre traffic congestion. Give **two** reasons for your choice. **(4 marks)**

5. How will the city of the future manage traffic issues? Use examples from cities today that might point the way ahead for a city like Edinburgh or Glasgow. **(6 marks)**

WHAT IS AN INTERDISCIPLINARY TASK?

Interdisciplinary tasks are when the knowledge and experience you have gained in a number of your school's departments comes together to solve a problem or to undertake a large-scale project. This will call on a whole load of the skills you have learned over the year and will involve several school departments working together. The task laid out below will give you an idea of how an integrated task might be set up. The specific task here may not suit your own school or your Geography department but it should be easy to change and adapt to your local circumstances if you want to get it going in your school.

THE BIG QUESTION: SHOULD SCOTTISH BATTLEFIELDS BE PRESERVED FOR THE NATION?

Background briefing: In the United States of America a large number of important battlefields have been bought for the nation and conserved as National Parks. The American government regards these sites as almost sacred ground and they are protected for ever from developments that might damage them or spoil them, as memorials to people who gave their lives for their country.

Battlefield	Location	Points of interest
Gettysburg June 1863	Pennsylvania	A large battlefield, the site of the most significant battle of the American Civil War. Huge numbers of visitors visit the battlefield and its visitor centre.
Antietam September 1862	Maryland	This was another major battle of the American Civil War and centred around a Shaker church and an historic bridge. There were 23 000 casualties on both sides.
Shiloh April 1862	Tennessee	A bloody civil war battle fought on the banks of the Tennessee River.
Little Bighorn June 1876	Montana	This battle is associated with General Custer's last stand. Originally seen as a heroic defeat for the US Cavalry. Recent historical evidence has changed people's view of the battle to something less glorious.
USS Arizona December 1941	Pearl Harbour Hawaii	This site commemorates those killed in the attack on Pearl Harbour by the Japanese, which brought the United States into the Second World War.
Yorktown 1781	Virginia	The last battle of the American War of Independence ended in defeat and surrender for the British, and the USA became an independent nation.

1. Look up one of these battlefields on the internet and discuss in a group why the Americans have preserved these historic battlefields and how they have gone about it.
2. Give your opinion: Should Scottish battlefields be specially protected from development? Give reasons why you hold this position.
3. Make a list of famous Scottish battlefields and create your own table like the one on page 112 but comment on the condition of the battlefield as in the example provided.

Scottish battlefield	Location	Comment
Stirling Bridge 11 Sept 1297	Causewayhead and Stirling	William Wallace's greatest victory, fought beneath Stirling Castle. Large areas of the battlefield are under housing, roads and other buildings. A significant part of the battlefield is still open space. All the land is privately owned.

AN INTERDISCIPLINARY APPROACH USING A LOCAL BATTLEFIELD

THE BATTLE OF METHVEN, JUNE 1306

This was a significant battle of the Scottish Wars of Independence. Robert the Bruce had been recently crowned King of Scotland at Scone near Perth. Edward I the English King sent an army north under the dragon banner (no mercy). His troops, under Aymer de Valance, occupied Perth and defended its walls. King Robert laid down the challenge for the English to meet him in battle and agreed to fight the battle the next day so as not to fight on Sunday as it was a holy day. Bruce's army camped a few miles from Perth beside a stream that runs through Methven Den. The Scots camp lies to the North of the present day village of Methven and the battle site is marked on the OS map of the area.

The Scots troops believing the battle with the English garrison to be a day away, relaxed at camp: knights took off their armour, men drank, cooked meals and relaxed without their weapons at hand and foraging parties went in pursuit of food. Only two-thirds of the Scots army were in camp when the attack began.

Persuaded by his aide Umphraville, that open battle against the experienced Scots army was a dangerous and unpredictable gamble, the English commander Aymer de Valance, instead organised a surprise attack. As the light fell, English knights supported by infantry fell upon Bruce's men. Warning horns sounded around the Scots' camp but it was too late. Some of the Scottish knights got to their horses but were unable to organise any defensive formations. Men on the ground were cut down and the battle became a running rout that continued for several days.

Although small groups of knights and infantry attempted to make a stand against the superior attacking force they were no match for troops acting in strength, formation and cohesion. Bruce himself was captured during the fighting but a Scottish knight fighting on the side of the English army (John de Halliburton) recognised the identity of his prisoner. When a suitable moment arrived he released the Scottish king who escaped through the nearby woods. Only 500 Scots escaped the slaughter and pursuit after the battle.

Battleground Perthshire, Two Thousand Years of Battles, Encounters and Skirmishes.
Paul Philippou and Rob Hands. Tippermuir Books, 2009.

THE BATTLE OF METHVEN PROJECT

If Scottish battlefields were to be preserved like those in the United States of America, how would you go about setting up a battlefield national park?

A secondary school near Methven in Perthshire has decided to look into this question and S1 students working in several departments in the school have worked with their teachers on this task. Work undertaken in a range of specialist departments will allow them to investigate the battle, and plan a visitor centre on the battle site. They will investigate a range of issues relating to planning the visitor centre and deciding on where it should be located. Below is a task list that shows what operations are being undertaken and which departments are making a contribution.

114

Department	Contribution	Design brief and operation in detail
Technology	Designing the visitor centre	In the Technology department, craft and design skills are used to design the Methven visitor centre building. This is the lead department in the project and they have laid out the design spec. for the students to follow. Students are tasked with planning and building a model of the visitor centre.
History	Historical research	As part of a unit on the Scottish Wars of Independence, students research the battle and its place in this part of their History course. A visit to the battlefield and to the Bannockburn visitor centre is part of their study.
Geography	Mapping and environmental assessment	In Geography, students have learned how to create maps from satellite images. They can also use OS maps and can measure using scale. They will use their geography skills to decide on the location of the battlefield centre. To do this they will research the likely impact of the building designed in Technology on the local environment, taking into account its effect on wildlife by doing an environmental assessment. They will also advise on the issues relating to visitor numbers, traffic management and road access.
Modern Studies	Village survey	In Modern Studies, students design a questionnaire to canvass the opinion of local people and to judge the impact of the visitor centre on the village of Methven. They will investigate how planning a new development is put through the local authority planning process. They have set up a class visit from a local authority planning officer. Investigate citizenship in relation to rights and responsibilities around access to the countryside.
Maths	Mathematical calculations	In Maths, students apply their mathematical skills to calculate and measure elements of the proposal. For example, by calculating the size of a parking bay in the school car park, they can work out the size that the visitor centre car park will be. Work on scale will be useful in map and model making.
English	Visitor interpretation	In English, students create text that informs visitors about the historical background and tells the story of the battle. PowerPoint presentations are made as audio visual presentations for tourists. Children role-play as tour guides and develop talks about the battle, the battlefield local wildlife and environmental issues.
Art	Visual display Board design	In Art, students design and produce display boards to accommodate the text produced in English. Striking pictures of medieval battle and heroic historic figures will be produced to form a display in the visitor centre.
Home Economics	Menu design	In Home Economics, students plan the menu for the visitor centre café and elements of the furnishings to be put into the space provided for refreshments and meals.
Drama	Drama production	Students study the play *Macbeth* and create their own drama about aspects of the battle with the help of the Drama teacher.

CO-OPERATIVE LEARNING: CLASS EXERCISES

Academic goals: to investigate a local battlefield to consider if it is worth preserving and to study the planning process involved in setting up a new tourist attraction, taking into account local interests and conservation issues.

Social goals: to discuss issues in teams and to make reasoned decisions that can be justified; team building.

 **PAIR–SHARE ICE BREAKER**

1. Form groups of four.
2. Taking turns for one minute, discuss with your shoulder partner these two questions:
   - Why are historic battlefields important?
   - Is the Methven Battlefield important and should it be preserved?
3. Next, tell your face partner what your shoulder partner said to you about this issue.

GROUP PROCESSING
Discuss how well you have managed the task so far and how you might do things better at the next stage. Decide on a view from your team.

WHOLE CLASS DISCUSSION
Each group should, in turn, express their point of view on the issue.

WINDOWS ON THE WORLD
1. Each co-operative group should draw a four-section window on a large sheet of paper (see fig. 10.2).
2. Imagine you are looking through the window at the battle of Methven as it is being fought. What would you see? Write it down in your window square. Stand up and work your way around the table reading everybody else's comments.
3. Turn the paper over and draw another window. Now imagine that the window is open. Write down in your window space what you would hear through the window as the battle was fought. Stand once again and read what others in your group have written.

116

STAND AND DELIVER

In turn, one person chosen from each group should tell the whole class what their group wrote: either what they saw or what they heard.

DEVELOPING A TEAM ACTIVITY

As a group, choose one of the departmental tasks in the list on page 115, and develop a plan of campaign that details how you will go about that part of the task. Prepare a presentation of your ideas to give to the class at a later date.

EXEMPLAR: GEOGRAPHY DEPARTMENT TASK

1. Use **mapping software** to get an OS map of the battlefield and nearby village.
2. Obtain a Google Earth **aerial photo** of the battle site.
3. Draw **a large-scale map** of the site based on the OS map and the aerial photo.
4. Mark on housing, roads, woodland, open space, water features and any points of interest that may affect the building project.
5. **Label the map** fully and add annotations that indicate points of interest to the planners.
6. Put on **scale and direction** indicators.
7. Deliver the map to the Technical Department for them to use in their building design.

Environmental assessment

Examine the aerial photos and the map resources. Decide what **wildlife habitats** are likely to be present in the area. Make a **list of animal and plant species** that could be affected by the building work. **Research** these plants and animals, find out if they are protected species and what would have to be done to prevent them being disturbed or displaced by the building work. Produce a **brief report** to the builders, noting any points they will have to consider before they start design and building work. Recommend whether or not the project should go ahead and give your reasons.

117

Roe Deer in Methven Den

Fox Near Methven

EVALUATING YOUR PROJECT WORK

It is important at the end of a large-scale project, particularly one that involves a large number of separate departments working each on their own part, to bring things all together and to evaluate the project to learn lessons for the future. What will the end product of your multidisciplinary project be? Will you produce a display for the whole school to see? Could you put on a presentation to parents at a parents' evening or on an open day? Could you arrange a competition like *Dragon's Den* or *The Apprentice* to present the best ideas? Maybe the school could present a project prize at prize giving and a display could be put on for parents at that event. Maybe you could write the project up for the *Times Education Supplement Scotland* or for your school newspaper.

 **CO-OPERATIVE TASK**

1. Get back into your project groups.
2. Taking turns at the table, each person should offer an idea they have about how the project display or event should be organised. Come to agreement as a group and put forward a proposal for the whole class.

DOTMOCRACY

Create a dotmocracy grid. Each person should vote for the presentation style they like best for the class event.

In groups of four discuss how the chosen event should be organised and be prepared to share these ideas with the rest of the class. Class discussion now should lead to the development of a class event plan and roles for everybody taking part should be allocated. Tasks should be set out, with everybody taking some of the responsibility for part of the event.

EVALUATING THE PROJECT

In groups of four, look back on the project. Evaluation tries to look at what was done well so that it can be used again in the future. Try to list as many of the positive elements of the project as you can and say why these things were successful and went well. Evaluation also looks at things that did not work out quite so well so that we can improve on them next time. Make a list of things that could be improved and say how the improvement might be achieved next time. Use the tables on the next page to record this for each group and then the class should discuss the evaluation among themselves.

What went well with the group activity?	Why did this work out?

What did not go so well?	How could this be improved on next time?

Alternatively, write out each of your group evaluations on a graffiti board. Post them on a wall and then do a gallery tour where each group reads every other group's posting before discussing the evaluations as a whole class.

OUTCOMES AND EXPERIENCES

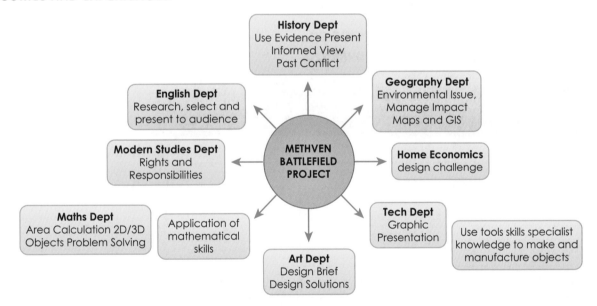

ANSWERS

GIS AND MAP WORK SKILLS PAGE 19

1. They are less detailed as they have no directions or map coordinates.

GLOBAL POSITIONING PAGE 20

1. The Equator, the Greenwich Meridian.
2. North and South.
3. East and West.
4. Zero degrees
5. Zero degrees
6. **B** 1 degrees south, 1 degree west, **C** 2 degrees north, 1 degree west, **D** 2 degrees north, 2 degrees east.

EXTRA TASKS PAGE 21

1. 30 minutes
2. 30 minutes
3. 1 degree 30 minutes south 1 degree 30 minutes west

CRASH RESCUE LOG PAGE 23

Ship 4 Island 3, Ship 5 Island 6, Ship 6 Island 48, Ship 7 Island 1
Ship 8 Island 10, Ship 9 Island 19, Ship 10 Island 42

1. There are too many islands to try.
2. One chance in fifty.
3. There are approximately 15 islands on the flight path.
4. One in fifteen
5. There are a limited number of islands in that direction.
6. They identify the square an island is in.
7. There may be more than one island in a square.

ANSWERS TO GET ACTIVE TASKS ON PAGE 24.
COORDINATES AND ORDNANCE SURVEY SYMBOLS

Ordnance survey map symbol	Coordinates on Map A
Windmill	G7
Coniferous forest	A7
Railway station	D3
Church with steeple	B4
A class road	D7
Motorway	G3

ANSWERS

GET ACTIVE SIX FIGURE REFERENCES PAGE 27

1. 228647
2. Railway Bridge
3.

Place	Four-figure grid references	Six-figure grid references
Kat Hall	1641	164413
Jean Halt	1842	181425
River Island	1741	174413
Spot height 5m	1841	185415
Twenty Acre Wood	1941	195415
Rabton farm	1739	178394
Monument	1537	153374
Jackton Bridge	1838	187383
Jackton Station	1939	192394
The quarry	1640	165406

ASSESSMENT 8 GET ACTIVE PAGE 30

1. Very steep slopes, rounded summits, hills run north west to south east, highest point 370 metres, lowest point sea level (zero metres).
2. It forms a large loop – a meander.
3. It has steep slopes that protect it from enemies; from this height, they would have been seen approaching. This location is also protected by the River Forth to the North.
4. It has plenty of space for building and expanding. It has attractive grounds. It has good road and rail links. It has a nearby town to provide workers, services and entertainment.

INVESTIGATING EARTH FORCES PAGE 32 ROCK TYPE CATEGORIES

1. Limestone – Sedimentary, Sandstone – Sedimentary, Basalt – Igneous, Marble – Metamorphic, Coal – Sedimentary, Quartz – Igneous, Gneiss – Metamorphic.

EARTH FORCES PAGE 47

1. Casualty figures tend to be estimates at first but become more accurate over time as more information is collected. Casualty figures always increase – early estimates tend to underestimate the scale of the event.
2. More people, with greater expertise, arrive on the scene, and reports of missing people come in.
3. People who are trapped need to be rescued quickly to secure their airways and to prevent death from dehydration or cold. Aftershocks may cause further deaths. Injured people need to be treated to prevent their injuries becoming more serious.
4. People are killed by collapsing masonry and buildings. Earthquakes can cause fires, gas leaks and explosions. Electrocutions and drownings might occur after tsunami events.
5. Britain is not close to plate boundaries.
6. The rule of three states that you can survive three minutes without air, three days without water and three weeks without food. Water supply is critical.
7. The Richter Scale records the magnitude of an earthquake according to the energy it releases.

ANSWERS

ASSESSMENT F PAGE 51

1.

A steep sided u-shaped valley left after the ice melts	C
V-shaped river valley	A
Glacier erodes the valley using plucking and abrasion	B

2. Glaciers occupy pre-existing v-shaped river valleys, which are re-shaped as the glaciers move downhill under gravity. Plucking explained (1). Abrasion explained (1). Diagram (1).

3. Each land use has specific reasons for being present – look for water supply and water features in lochs **OR** look for the influence of spectacular scenery **OR** look for opportunities for sports and outdoor activities. Other land uses, such as growing crops, may be uneconomic so forestry and hunting make use of land that has few other potential uses. Military training prefers remote unoccupied areas for safety and security. Energy needs steep slopes to generate hydro power. Remote areas can be dammed without displacing people.

ASSESSMENT G PAGE 57

1. Large meander – 815944 or any other large meander. Flat floodplain – any square along the Forth with no contours. Flat slope with v-shaped valley – any river and valley from the north east corner of the map.

2. The river has a series of wide meanders e.g. 803943 (1). The meanders are incised or tight, closing on themselves to form ox-bows (1). The river is tidal from Old Mills Fm 779957 (1).There is a wide floodplain alongside it. There are embankments for flood protection.

3. Ox-bow: Erosion takes place using hydraulic action and corrasion (1). The outside of the meander is eroded (1). Sediment is left on the inside of the bends deposition (1). The insides of the bends are eroded narrowing the gap between (1). The river breaks through and seals the ends of the ox-bow with sediment setting up a new course (1). Diagram (1).
   Waterfall: The alternating beds of hard and soft rock are eroded at different rates (1) by hydraulic action (1) and corrasion (1). The weaker beds are undercut (1) and this causes the harder beds to collapse (1). The eroded rock swirls and deepens the plunge pool beneath the waterfall (1). The process repeats itself, forming a gorge (1). Diagram (1).

ASSESSMENT H PAGE 61

One mark per valid point

1. Hydraulic action, corrasion, corrosion, abrasion (1). Undercutting the cliff, wave power and compressed air (1). Rocks knock bits off the base of the cliff and undercut it (1).
2. Softer rocks are easily eroded. (1)
3. Cave, arch, stack, blow hole. (1)
4. Spits and bars or tombolos. Beaches and sand dunes. (1)
5. They are good places for wildlife such as wading birds. (1)
6. Explain by referring to weaknesses such as faults (1) and weak rock beds (1). Include erosion terms given in answer 1 above. Show how caves (1) become arches (1) and arches become stacks (1). Diagram (1).

ANSWERS

INVESTIGATING CLIMATE PAGE 64

1. 3 Degrees Celsius in January, 15 Degrees in August
2. 12 Degrees Celsius
3. Approx. 660 mm
4. Highest in July, lowest in February, March and April
5. The highest temperature matches the highest precipitation (additional convectional rain in summer). The lowest precipitation matches the lowest temperatures. (Cold air is often less moist and less likely to deliver rainfall.)
6. Starting in January, the temperature is at its lowest (2 degrees Celsius) rising only slightly in February, but then rising steadily to reach a peak of 14.5 degrees Celsius in July and August. In September the temperature declines steadily and then declines yet further to reach 6.5 degrees Celsius in November and 4.5 in December.
7. In a similar way to the answer above note the upward and downward trends and support with figures from the graph in the description.
8. July as it has the heaviest precipitation.
9. Any of the months that have the lowest precipitation.
10. Summer gets the most precipitation because not only is there the normal rain associated with depressions and fronts crossing Scotland, but there is additional rainfall from thunderstorms and convectional uplift that produces heavy summer showers.

ASSESSMENT I PAGE 70

1. Wind SW 25 kts, –5 degrees, rain, full cloud cover (8 oktas)
   Wind NW 5 kts, –2 degrees, snow showers, cloud cover 4 oktas
   Wind E 1-2 kts, 20 degrees, thunder storms, cloud cover 6 oktas.

2.

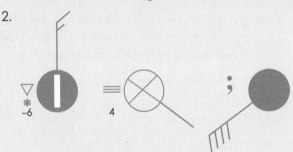

3. a. Wind NW 15 kts, –2 degrees, 8 oktas, b. Wind 5 kts SSW, 3 degrees, 2 oktas, e. Wind NW 25 kts, 2 degrees, 8 oktas, g. Wind W 20 kts, 6 degrees, hail showers 4 oktas. At station h. Winds blowing at 25 kts from the SW raining, 6 degrees, full cloud cover. There is a warm front passing over and this brings persistent rain as the warm air in the system is forced to rise over the cold air in front of it.

PYRAMID POWER PAGE 99

1. Brazil – Wide at base
2. Scotland – Wide at top
3. Brazil because it has large numbers of young people
4. Scotland because there are more people at the top of the pyramid
5. Brazil because it has a larger overall population. Proportionately Scotland has a bigger working population.
6. Brazil because of the large numbers of people in the reproductive age groups and the large number of youngsters.
7. Scotland as it has a lower birth rate than Brazil but also because it has traditionally provided many emigrants to other parts of the world.
8. Brazil because it is a ELDC and it has a growing population. It will need to develop its resources quickly.

ANSWERS

CO-OPERATIVE LEARNING

Social skills: to compete with others on a task; to discuss a topic effectively by taking turns.

Learning intention: to apply knowledge of latitude and longitude to a map problem. One way to play the Space Rescue game is as a whole class, with groups of four students competing against each other to rescue as many ships as possible. Set up co-operative teams.

1. Work together to complete the crash ship rescue log and then compare results with other groups in the class.
2. Use the 'Around the table' method to complete the game questions:
- Number each group member 1 to 4.
- Starting with person number 1, each person gives their answer to question 1.

- It is important to listen to everyone's point of view but it is also good to disagree, provided you give good reasons for doing so.
- Discuss the group answers and come to an agreement.
- Write out your agreed group answer.
- Move on to the next question, starting this time with person number 2 and repeat the procedure.
- Answer all the questions in turn.
- Finally, taking turns around the table, say why you think using latitude and longitude is a good way to find places on maps.

FOLLOW UP

1. Use the index of an atlas to look up places. Write down the page number and the latitude and longitude coordinates provided and then find the places in the atlas using latitude and longitude.
2. How many cities can you find called Glasgow, Perth, Aberdeen and Edinburgh in other countries? Find them using latitude and longitude coordinates in the atlas.

3. Does your own town share its name with other places in other countries? Make a list like the one in question 2 for your home town.

124

CONTENTS

INTRODUCTION

There are few cars more enduring than the Volkswagen Beetle. It began life as Ferdinand Porsche's concept of a small car for the man in the street - a concept which suffered from innumerable false starts. The wartime influences on its development could so easily have branded it with a completely negative image, an image from which sales would never have recovered. But the car survived and prospered. Building it provided work and shelter for thousands of displaced people after the war, and the town of Wolfsburg owes its existence entirely to the Volkswagen.

Beetle sales eventually outstripped that of the Model T Ford, and it is still being produced today in Mexico and Brazil. Amazingly, although just about every small part of it has been changed in some way over the years, the very youngest of Beetles still looks just like the oldest - an instantly recognisable automotive legend throughout the world.

Andrea Sparrow

PORSCHE
& THE BEETLE

Ferdinand Porsche

Ferdinand Porsche was born in September 1875 in the town of Maffersdorf. By the time he was fifteen years old, he was busy in his father's tinsmith's workshop and spending every moment he could studying engineering - formally at the local college in the evenings and, informally, by observing everything that was going on around him. This was a time of great industrial change. Ferdinand was fascinated by electricity, the new technology of the moment; his natural talent for engineering opened his mind to the great possibilities that it offered. He was soon earning extra cash by making and fitting electric doorbells, as well as building and installing an electric light system for his father's business. The young Porsche was also intrigued by another novelty - the internal combustion engine. A local businessmen was the proud owner of one of Herr Daimler's very first motor cars.

At the age of twenty-one, Porsche moved to Vienna to work for an electricity company. He quickly established a reputation both for brilliance and for hard work. He still studied whenever possible; he attended lectures at Vienna technical college and at the University although, if he was spotted, he was thrown out as he had no right to be there at all! Vienna was a thriving city attracting people with wealth and connections so motor cars, the latest talking point, were often to be seen there. Porsche's interest in them was growing all the time; he began to give serious thought to the engineering problems faced by automobile designers and to what the solutions might be. Then, in 1898, the company Porsche was working for was approached by Ludwig Lohner, who was having some problems with the electrical motors he was developing to power his automobiles. Porsche was called upon to provide some answers. Lohner was impressed, and offered Porsche a job as his chief designer which, naturally, Porsche accepted.

Porsche stayed with Lohner until 1906. During his time with the company, he developed an electric hub motor to drive the automobiles he was designing. Electricity for the hub motor was generated by a petrol engine (such as that made by Daimler). Lohner could not afford to spend vast sums on innovation, so a frustrated Porsche joined the Austrian Daimler company (which no longer had ties with the German company) as a director. His new company needed a good shake-up, particularly in the organisation and employee-relations departments. Porsche was given a good budget to work with, and set about the task enthusiastically.

Almost immediately the result was a happier workforce, a much more productive and profitable company. After this spectacular start, Porsche settled down to revitalise the company's products, including the development of racing cars which would win many awards, often with Porsche himself at the wheel.

By the time the first world war was over, Porsche had risen to the rank of managing director. He was also astute enough to know that the climate had changed in the automobile world. Austria was

then, and for the foreseeable future, a poor country. Porsche calculated that what was needed was a small economic car; a car that would sell to the people. But his colleagues disagreed. Smaller cars were appearing elsewhere in Europe, but Porsche could not persuade his co-directors at all. In early 1923 matters came to a head and Porsche resigned.

Porsche moved to Stuttgart, to take up the post of technical director at the German Daimler company. At first he met some resistance,

Made in Germany in 1959 - approximately half-way through the Beetle's production life there.

The unmistakable and timeless shapes of the Beetle.

Overleaf - There are many international clubs dedicated solely to the much-loved Karmann convertible Beetle.

The Beetle has always been a car for people of all ages - and especially people in love.

Monument to a great idea.

but soon made his mark. He revitalised the racing division of Daimler, and went on to develop a series of cars bearing the Mercedes badge; a series which included the famous SSK. Porsche managed to whet the appetite of his co-directors for his small car project. By 1927, prototypes had been built and tested, and a series of 30 commissioned for further testing. Unfortunately, Daimler had recently absorbed the Benz company, whose former directors tended to oppose Porsche on principle. They managed to get the small car project shelved. Porsche was furious, and handed in his resignation.

Porsche next joined Steyr, in Austria, where he developed several highly thought-of and commercially successful vehicles. However, fate intervened in the form of the collapse of the *Österreichische Bodenkreditanstalt*, a bank which owned a major share of Steyr. Fortunately for Steyr, the bank was rescued by another, the *Kreditanstalt am Hof*. Unluckily for Porsche, this new bank controlled Austro-Daimler, so it would have been inevitable that Porsche would, once again, have to work with the very people he had fallen out with before. So he decided to break with Steyr and returned to Stuttgart.

By 1930, Porsche was established independently at premises in Stuttgart. Among those working for him were his son Ferry and Joseph Kales, who had been involved in the design of the air-cooled engines used in the Czech Tatra cars. Thoughts of the small car were still never far from Porsche's mind, and he made many notes and sketches as he thought through the various design possibilities. Of course, he would have to find a company that would be prepared to back the project and build the car. Porsche set his team

to work on the nuts and bolts of the design. He was a hard master, questioning and revising endlessly until he was satisfied. By the end of 1931, the design was complete; a rounded shape, with two doors, a three cylinder air-cooled engine at the rear and independent suspension on all four wheels.

Porsche tried to drum up interest in the industry for his small car project, with no success at first. Then Porsche was approached by Dr Fritz Neumeyer of the Zundapp motorcycle company. Neumeyer had heard rumours of Porsche's small car - it sounded as if this might fit in very well with his own plans to build a people's motorcar - one which the man in the street could afford. The two men met, and soon came to an agreement. Neumeyer would share some of the development costs, and three prototypes would be built immediately. Neumeyer stipulated that the engine should be water-cooled and of five-cylinder design: Porsche was happy to agree, as it meant seeing his long-held dream going into production at last. The prototype

Air-power and wind resistance! The Beetle's flat windscreen was always a deterrent to high speed.

A Beetle enjoys a camping holiday.

bodies were produced by the firm of Reutter, and whole cars were soon assembled and being tested. But there were serious problems. The cooling was completely inadequate, the gearbox was riddled with faults and the suspension broke. Porsche persevered, but the general economic situation was worsening fast so Neumeyer decided, reluctantly, to abandon his small car project.

Porsche was back at square one, but did not give in. He was soon talking to NSU about a small car of the same basic shape as before, but this time with a four-cylinder air-cooled engine and completely redesigned transmission and chassis. Three prototypes were built, two of them bodied by Reutter. There were problems, but they were solvable and the cars performed well under a variety of conditions. Then, again, it all went wrong for Porsche. Fiat reminded NSU that the two companies had an agreement. When Fiat had bought NSU's previous car-making facility the terms were that NSU were forbidden to make cars again. The agreement was written in stone. Porsche's feelings can well be imagined, but it was not in his nature to give up. He put the project to one side, confident

Opposite page - A wonderful Polar Silver Beetle from 1956 and ...

... an equally impressive 1951 model, both cars fitted with lots of period accessories.

that, in a happier economic climate, its day would come.

The Beetle's Birth
One day, in 1933, Jakob Werlin was in Stuttgart on a short business trip. While there he decided to pay a visit to one of his colleagues from earlier days at Daimler - Ferdinand Porsche. Werlin had done well at Daimler, and was in charge of Mercedes sales

for the Munich area. He'd also become a friend and associate of Adolf Hitler; many times the two men had discussed Hitler's plans to encourage the making of smaller and cheaper cars in order to give every German the chance to own a motor car. Talking to Porsche, Werlin realised that here was someone who had already gone some way down the road towards a 'people's

A 1951 export model Beetle. Although Porsche admired Henry Ford - whose Model T could be had in any colour so long as it was black - Nordhoff decided that black should be just one of several colour options for the Beetle.

The 1302. MacPherson strut front suspension was a radical change and bodies were much more prone to rusting than earlier cars.

car,' and who had the necessary credentials and expertise to see such a project through.

Porsche was summoned to Berlin to meet Werlin again. He was surprised, when he arrived, to be told by Werlin that he was to meet Hitler, who wanted to discuss with him his plans for a small car. Porsche had met Hitler before, but had no idea that Hitler remembered him. By the time Porsche left Berlin, he knew what Hitler had in mind; a small air-cooled car to take two adults and three children and their luggage, cheap to run, suitable for the new autobahns - and all for a price of under 1000 Marks.

Porsche worked on designs, costings and related commercial decisions for the project and, by the beginning of 1934, he'd got far enough to submit a plan to Hitler for the production of a *Volkswagen* (Peoples' Car). He was careful not to commit himself to the target price, as he was far from sure that the car

could be sold for as little as 1000 Marks - privately he believed that 2000 Marks would be hard to achieve.

Two months later, the 1934 Berlin Motor Show was opened by Hitler with a speech in which he said he regretted the lack of affordable cars for the ordinary person. After the show, Porsche was introduced to the transport minister, who told him that the idea of the Volkswagen project had met with approval - but he should go away and work on the cost - 1000 Marks maximum! By

These 'flat' hubcaps replaced the classic domes in 1965.

June though, Porsche had received the news that the Volkswagen project had the green light. Various manufacturers would be instructed to sell him parts at reduced prices. He was to start work on the first of three prototypes, which should be ready in ten months time.

Initial work on these cars went well, but there were the usual prototype problems to overcome. However, Werlin told Porsche that Hitler was not so worried about timescale as might be assumed. Towards the end of 1936, Porsche visited the USA to observe the mass-production methods of the car industry there; he was well aware that, for the price constraints he had been given, traditional European production methods would not do.

In October 1936, the three prototypes were officially presented to a representative of the Society of German Automobile Manufacturers. Exhaustive tests were begun, covering all types of road surface and condition. By December, each of the three cars had covered the requisite 50,000 kilometres. The officials responsible for analysing the tests gave their opinions; the Volkswagen project could succeed and merited perseverance.

The Volkswagen company was founded under the watchful eye of Robert Ley,

Head of the German Labour Front, who was in charge of, amongst other things the *Kraft-durch-Freude* (Strength through Joy) movement - KdF for short. Porsche was given the go-ahead to make whatever technical improvements and alterations he felt necessary, bearing in mind the brief he had been given. Money was not a problem, as Ley was organising whatever finance Porsche needed. Werlin arranged for a batch of 30 more prototypes to be made, this time by Daimler-Benz. These cars, known as "Series 30," were ready for action within four months, and immediately embarked on a strin-

12-volt electrics made the outlook a lot brighter for the Beetle.

what he could out of the parts that were available. There was a huge job to be done, clearing and rebuilding. There was plenty of labour - thousands were fleeing the Russian sector - but this brought a new set of problems, as the new arrivals had to be housed and fed.

Volkswagen, the concept, the car and the plant, came under careful scrutiny from a host of British experts at this time - and their reactions were mixed. The British authorities would have been happy to sell the lot - but there were no offers. They were reluctant to shut it down, partly because it was useful to have a source of vehicles for use in Germany, and partly because it solved the local unemployment problem, which helped keep the community stable. However, they were also aware that the plant would, in the end, need to be handed over to the German government, so they searched for someone to be trained to head up the whole operation. They chose Dr Heinz Nordhoff, who had worked for BMW and Opel. He joined the company at the start of 1948, becoming managing director in September 1949, when the British authorities relinquished control of the plant. The task facing Dr Nordhoff was a colossal one. He needed machinery, some of it from abroad, and the only way of buying it was to use foreign currency earned through exports. Although he managed to sell almost 40,000 cars in Germany during 1949, thanks to the now stabilising conditions, he implemented a huge export drive. He backed

up his product, sometimes unpopular just because it was German, with a first class servicing and parts system. His plans worked. In 1950, he doubled the factory's output and quadrupled export figures.

Dr Porsche died at Stutt-

gart in January 1951. His small car project was being realised at last by Dr Nordhoff and the town of Wolfsburg was growing and providing itself with the facilities it needed, often with the assistance of the Volkswagen company. Peter Koller, the town's original

The few saloon cars that were produced during the war were used by high ranking Nazi party officials.

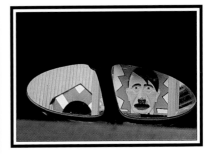

Hitler had well-appreciated the political capital that could be made out of Porsche's idea for a peoples' car.

architect, who had fought as an ordinary soldier and spent two years in a Russian prisoner-of-war camp, returned to Wolfsburg, and was commis- sioned by the newly formed town council to build one of the two churches that he had always known should be at its heart.

EVOLUTION 3

Throughout its extraordinarily long production life, the basic shape of the Beetle has hardly changed. The changes that did occur were updatings and improvements rather than radical redesigns - evolution rather than revolution. Changes came about as a result of improved technology (the switch to 12-volt electrics and increases in horsepower), or as a response to changing driving conditions (bigger rear window and bigger rear lights) or, occasionally, to fit in with a need for a more modern style (a cleaner-looking dash panel).

The Beetle that finally went into production in 1945 was fitted with the 1131cc (bore and stroke of 75mm and 64mm) engine used in the *Schwimmwagen* - so nominally the Beetle was a 1200cc car. The engine followed Kales' Tatra design, being a four-cylinder horizontally-opposed 'boxer' motor, cooled by forced air which was delivered via an upright fan sharing the spindle of the belt-driven dynamo bolted to its own mounting on the crankcase. Fins on the cylinder barrels and aluminium alloy cylinder heads (two barrels per head) created a greater surface area and thus more efficient cooling. The barrels were bolted to a two-piece crankcase, which was also the oil container, there being no separate sump as such.

The crankshaft was of forged steel and centrally mounted. Directly above it, and driven by direct gears, sat the camshaft, which operated the valves via pushrods, each running in its own separate tube. The distributor was placed at one end of the crankshaft, from which it was gear driven and was (and still is) easy to reach for routine servicing.

With a compression ratio of 5.8:1, the motor developed 25bhp at 3300rpm. Originally the engine used a Solex downdraught carburettor, supplied with fuel via a mechanical pump. (At the end of the war, Germany was segmented into four occupation zones; the Solex factory, being in Berlin, fell in the Russian zone. Supplies of Solex carburettors dried up until 1950 so, until then, a carburettor with a die cast aluminium body and float chamber was manufac-

July 1974. It was number 11,916,519. Beetles were still being produced elsewhere though - at Emden in Germany and Brussels in Belgium, in Brazil and Mexico, and at other, smaller plants throughout the world. The last German Beetle - number 16,255,500 - was produced at Emden in January 1978. From then on, the Beetles of Europe came courtesy of the Volkswagen factory in Mexico. It was in Brazil that the 20 millionth Beetle was produced in May 1981. The Beetle's amazing production total had seen the 1 million mark passed in August 1955, 5 million in December 1961, 10 million in September 1965 and 15 million in February 1972.

The last Beetles had the front indicators incorporated into the bumpers, leaving the wing tops unadorned as they were on the first cars.

Previous page - From 1978, the Beetles of Europe come from Mexico, including this special edition Jubilee Beetle in gleaming silver. Ole!

UNUSUAL
BEETLES
4

The 1970s saw a craze for special edition cars gaining momentum. In 1972 Volkswagen produced three special editions of the Beetle. Marking the 30,000th British example off the production line was the GT Beetle, a limited edition of 2500. It was based on the 1300, but came with the 1600 engine and sports wheels. The Marathon, which celebrated the Beetle's breaking of the Model T Ford production record of 15,007,033 units, and a Summer special - the June Bug.

Other special editions included the Jeans Beetle in bright yellow/orange with stitch-look paintwork embellishments and denim seat covering, the Chocolate Beetle, a popular 1303 special in brown with beige trim. The 'Sun Bug' came in two colour

Not that many people would not recognise this as a Volkswagen - but this streamlined roundel accessory confirms it.

A beautiful 1956 Beetle
owned by Mon Martens
fitted with a whole host of
period accessories both
inside and out.

THE VW FAMILY

It was understandable that Nordhoff should initially want to concentrate on one model of Volkswagen. Its potential was obvious to him, and that potential was beginning to be realised. But, even as early as 1947, the company was considering a van based on the Beetle. A prototype had been built and tested in 1949 and, in March 1950, the Type 2 Transporter was introduced. Its concept was simple; whereas the Beetle was made for carrying people, the Transporter was designed for loads. Based on the Beetle's mechanicals, the Type 2 began as just a van, but soon buses, utility vehicles, ambulances, pick-up trucks and campers appeared. The 'bus' hit the heights of popularity in the 1960s as a camper and has recently appeared in psychedelic livery on poster advertising for the VW people

The Type 3 - not one of VW's greatest success stories - but it was popular enough to stay in production for twelve years.

The Type 2 has carved its own place in the history books; versatility being one of its strengths. This early example has been an ambulance first for a Belgian hospital and then in a private capacity.

carrier - 90's style.

Beetle production had passed the five million mark when another new model was introduced in 1961. This was the Type 3, although at its launch it was simply billed as the VW 1500. It followed the same basic principles as the Beetle, with separate chassis, air-cooled engine at the rear and torsion bar suspension. Unlike the Beetle's though, the engine cooling fan was placed horizontally, which gave the Type 3 an advantage over its older sibling; it had luggage space in the back as well as the front. Ride quality, handling and interior comfort had been improved, with the consequence that The Type 3 was also considerably more expensive. In addition to the original saloon version, there was a roomy estate, with the Variant fastback appearing the following year. It had been intended to produce a cabriolet version, but it never went into production, as the strengthening involved would have made it much too expensive to be commercially viable. The Type 3 never even approached the Beetle's success, although it remained in production until 1973, with just over two million having been manufactured in all.

In 1968, VW announced another new model, the Type 4. The first model, the 411, which was available in either two or four door versions, was powered by a 1697cc air-cooled engine mounted similarly to that of the Type 3. There was also a 1795cc engined version, the 412. Both

A 1978 American specification Type 2 Camper van - a Westphalia conversion with two-litre fuel-injected engine.

The Type 3 Variant in an unusual but practical guise and ready for fire drill.

Arguably, the nicest-looking of the Type 3s - the fastback which appeared a year after the other versions.

An advantage of the Type 3's horizontal engine cooling fan was the provision of luggage space in the rear compartment.

The first Type 4 - the 411. More cumbersome than its predecessor and, unfortunately, with inferior handling.

cars came in estate versions too. The Type 4 was never popular, due in part to its wallowing suspension and tendency to break away at the back when cornering hard on wet roads. Only 400,000 rolled off the production line.

There was another, more unusual, Beetle-based car introduced in 1969 to the USA market under the name of The Thing. Its official designation was Type 181, and it was built on the slightly wider Karmann-Ghia chassis. The UK version, Type 182, was named Trekker, although only 300 were ever imported. Production

The 412 in estate form - although never hugely popular, it did offer lots of load carrying space - but 'cornering on rails' was not its forte.

GS·FZ·23
VW 412 CLUB HOLLAND

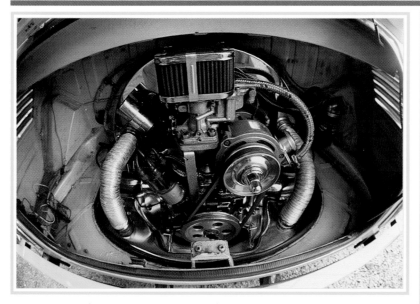

A familiar sight in unfamiliar surroundings.

switched in 1973 from Germany to Mexico. Although it was designed as a fun vehicle, the Trekker/Thing was very much in the Kubelwagen mould, and several were sold for military purposes.

In 1969, Volkswagen took over the NSU company. NSU had developed a new front-wheel drive car with a water-cooled engine, but the launch was postponed by the takeover. Thereafter, the car, the K70, was launched by Volkswagen under their own name. In itself the K70 was not very popular - just over 200,000 were built in all over five years, but it did set Volkswagen on course for a future that would bring success stories such as the Passat, Golf and Polo.

came a run of 25 cars for extensive testing; the results proved to all concerned that the Karmann Beetle cabriolet could be a successful model.

At the same time that he gave instructions to Karmann, Nordhoff also commissioned the firm of Josef Hebmuller and Son to work on a convertible version of the Beetle. Hebmuller solved the strengthening problem in a different way to Karmann - with two Z-shaped reinforcements under the car and a gutsier windscreen surround. The Hebmuller was a two-seater, with a distinctive engine cover that looked like the front bonnet, giving it a symmetrical look: classy two-tone paintwork gave a finishing touch. The prototypes were tested, and, as with the Karmann convertibles, Nordhoff was satisfied with the results. He commissioned a first batch of 2000 cars from both Karmann and Hebmuller, asking that original Volkswagen components should be used whenever practically possible. Both cars were launched together in 1949 as alternatives - a two-seater and a four-seater to complement the saloon Beetle.

For the Karmann, production began to gather pace, another 2000 cars being ordered within six months. Because Karmann was basically buying cars from Wolfsburg and converting them, and bearing in mind his brief

An extrovert Karmann Cabriolet wearing elegant spats over the rear wheels.

to stick to Volkswagen specification, it's not surprising that most of the production changes to the saloon car over the years appeared on the convertibles too. In 1972, the Karmann convertible was given a completely new hood with steel frame; an enlarged rear window echoed changes to the saloon car. Production of the convertible continued into 1980, with well over 300,000 produced in all.

As for the Hubmuller, it never did fare as well. It was well received, but a serious fire at the factory soon after production had started ruined its chances. Production slowed, and less than 700 were ever built. Hebmuller went into liquidation in 1952, and Karmann took over the production of the last few Hebmuller convertibles, the very last leaving the factory in February 1953.

Wilhelm Karmann had built his reputation on quality products for other manufacturers, and the Beetle convert- ible confirmed and strengthened this reputation. But he had always wanted to build a Karmann-badged sports coupé of his own, an idea that he and his son - also called Wilhelm, who took over the company on his father's death in 1951 - discussed both with their business associates in Wolfsburg, and separately with Luigi Segre at Ghia in Turin. In 1953, Segre showed the younger Karmann his new prototype project - a coupé based on the Beetle which

Hood up ...

The Hebmuller Cabriolet still has a following. Steel panel kits, like this one from Australia, will fit any Beetle shell. Kits are available from specialists like Wizard in the UK.

... or hood down, the VW Cabriolet is always a class act.

realised almost exactly the car which Karmann senior had envisaged. Segre had procured his Beetle through the French importers, as Dr Nordhoff had refused to supply one. But Dr Nordhoff soon got the chance to inspect the prototype and, aside from a few minor alterations, gave it his blessing. The Karmann Ghia Coupé was launched in July 1955 to a mild reception, but after its appearance at the Frankfurt show in September, interest gathered momentum and production increased. The cars were expensive to produce due to the complex shape of the panels, which were largely formed by hand.

The new Karmann Ghia was soon surrounded by controversy, as the Chrysler Corporation claimed it was a scaled down dead ringer for prototypes that Ghia were working on for them, from designs by Chrysler's own designer Virgil Exner.

In August 1957 a convertible version, which had been on the cards since the building of a prototype three years earlier, went into production alongside the coupé.

Karmann also produced a car based on the VW type 3 which, although stylish, never became really popular. In the seven years from 1962 when it was introduced, only just over 42,000 were produced.

Karmann was also responsible for the production of the VW-Porsche 914, the mid-engined sportscar that was powered by either a VW or a Porsche-built engine. The size of the Karmann operation made them the ideal choice to produce the car, their mass-production methods and traditional hands-on craftsmanship working side by side. This happy mix has kept Karmann at the forefront of international coachbuilding, producing nowadays vehicles such as well-appointed Camper/MPVs based on Volkswagen light commercial vehicles.

American laws demanded side marker lights which rather spoil the Karmann Coupe's excellent lines.

This two-tone Coupe is unrestored and completely original - perfect in every detail.

The Karmann company specialised in convertibles - so you would expect the softtop version of their own car to look good - and it does.

This Karmann Coupe was based on the VW Type 3 chassis. Though stylish, it fared no better than the Type 3 itself.

The rear of Karmann's Type 3-based Coupe - more angular than the Beetle-based version.

GALLERY 7

Karmann Cabriolet hood folds neatly and easily.

Supple torsion bar suspension copes with any surface, even cobbles like these.

Always chic, the Cabriolet has plenty of luggage room for a couple travelling in style; families would have to resort to luggage racks on their practical saloons which was not a problem as these racks were available from the start.